# GALAXIES
# COLLIDE

# GALAXIES COLLIDE

## AARON L BRATCHER

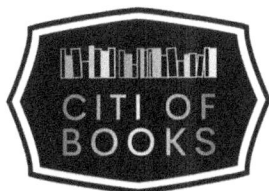

CITI OF
BOOKS

**CITIOFBOOKS, INC.**
3736 Eubank NE Suite A1
Albuquerque, NM 87111-3579
*www.citiofbooks.com*
Hotline: 1 (877) 389-2759
Fax: 1 (505) 930-7244

Ordering Information:
Quantity sales. Special discounts are available on quantity purchases by corporations, associations, and others. For details, contact the publisher at the address above.

Printed in the United States of America.

| ISBN-13: | Softcover | 978-1-963209-70-9 |
| --- | --- | --- |
| | eBook | 978-1-963209-71-6 |

Library of Congress Control Number: to be followed
No ChatBots were used in the creation of this story.

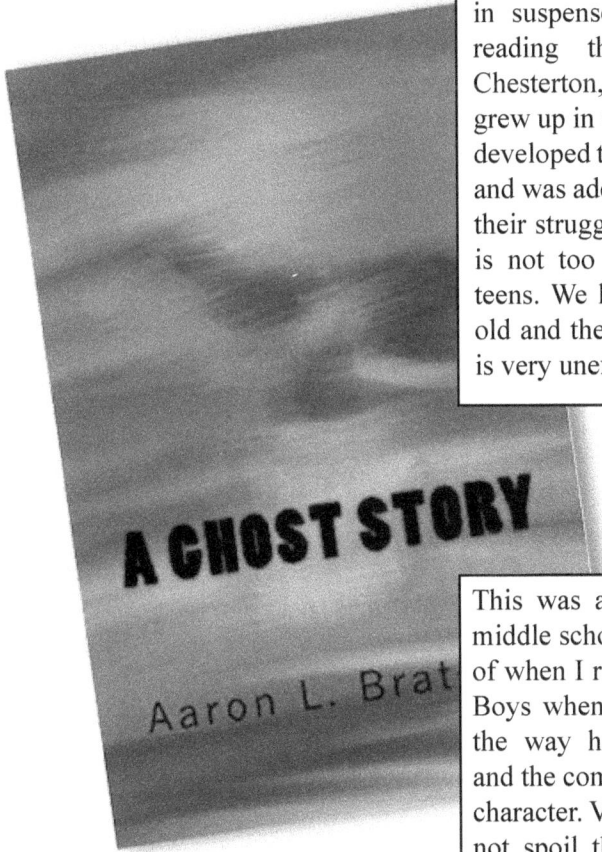

This is a great read for young adults/teenagers. We read it on a long road-trip and everyone was in suspense. We really enjoyed reading the parts about the Chesterton, Indiana area since I grew up in Chesterton. The author developed the characters very well and was adept at engaging us with their struggles as teens. The story is not too scary for young pre-teens. We have a 13 and 11 year old and they loved it. The ending is very unexpected.

A GHOST STORY

Aaron L. Brat

This was a very good book for middle schooler's. It reminded me of when I read the original Hardy Boys when I was young. I liked the way he mixed the mystery and the coming of age of the main character. Very entertaining! I will not spoil the ending, but it was total surprise, and I remember saying out loud, "What ???"

*Get your free e-book!*

https://AaronLBratcher.com

# Table of Contents

# Landing

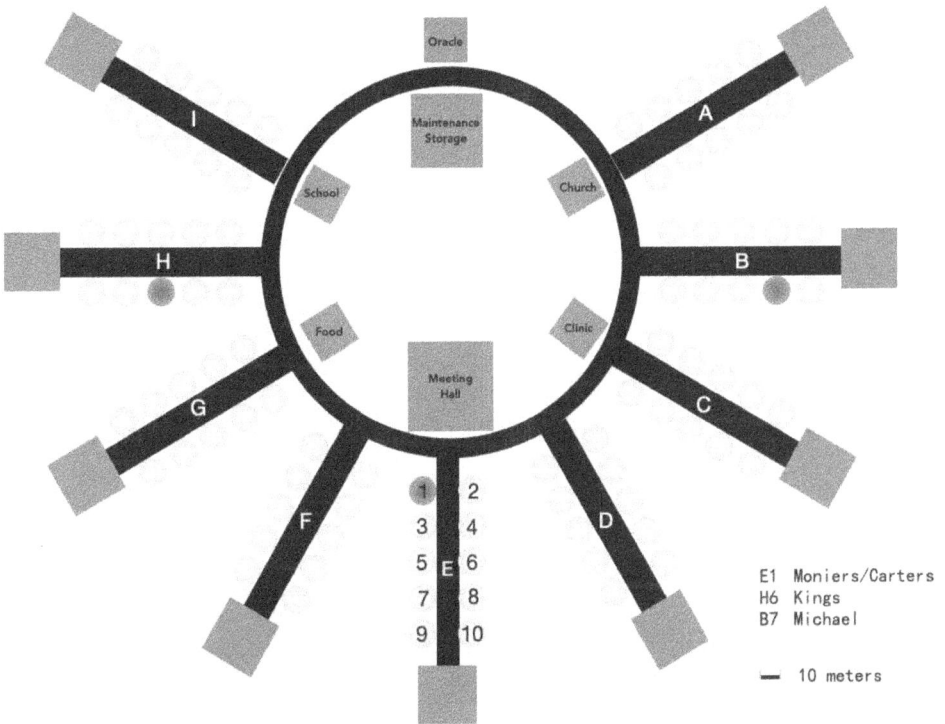

Oracle

Maintenance
Storage

I

A

School

Church

H

B

Food

Clinic

G

C

Meeting
Hall

F

D

| 1 | 2 |
| 3 | 4 |
| 5 | 6 |
| 7 | 8 |
| 9 | 10 |

E

E1 Moniers/Carters
H6 Kings
B7 Michael

— 10 meters

# Prologue

Dr. James Monier watched the vertical green bars vibrate on the display, the tops flashing hints of orange. Speaking to nobody in particular, he muttered, "This storm is making me nervous." He glanced up to the wall where a timer counted down, showing one hour and forty-eight minutes remaining. Technicians sat nearby in the white ten-by-twelve-meter room, their eyes on various colorful displays. A window dominated most of one wall, overlooking an underground chamber large enough to hold a football stadium. At the end, below and just to the right of Dr. Monier, a sheet of greyness shimmered with flashing pinpoints of color.

Movement on the floor caught his attention. A woman dressed in a bright red polyester shirt and black pants strode up to and faced the shimmering wall.

"What is she doing?" he whispered to himself. Without saying a word to anyone in the room, he dashed to the elevator. Once on the chamber floor, he called out, "Amira?" His voice sounded strangely muted, the wormhole affecting the sounds in the area. He ran up to her, put his hand on her shoulder, and leaned close to her ear. "Amira?"

She turned, her eyes bright with excitement. "James! Come to see me off?"

"What are you doing?"

"I think that's plain to see—I'm going through the wormhole. Just a quick step through to see it for myself." She held his eyes. "We know it's safe. How many mice do we need to send through before we can go ourselves? I want to be first. Then I'm going to spend time with Fanar. You should come home with me when I get back. It's been a while since you spent some time with our son."

"I have a lot of work to do. He's fifteen; I'm sure he'll understand." Her eyebrows drew together in concern. "James, he's sixteen."

The statement weighed on him. *Sixteen? Has it already been another year?* "He won't want to see me. I've already lost him."

Amira cupped her hand against his cheek and leaned in close. "He loves you. You haven't lost him yet." She turned to the shimmering wall and looked over her shoulder, a challenging gleam in her eye. She raised her voice. "But first, I'm going to Niton."

He smirked and shook his head. How could he deny his wife? He pointed at a large countdown timer on the wall, in sync with the one in the control room. "How long?"

She held up a water bottle. "Just long enough to drink this." She blew him a kiss and, before he could say anything else, disappeared into the grey. The entire wall flashed white just as Amira disappeared, causing him to flinch and close his eyes. Startled, Dr. Monier tapped the uplink on his wrist and yelled into it, "What happened?" A muted voice responded, "A lightning strike on the power line. Just a small spike. Everything's good."

Slowly nodding, he slapped the uplink and stared at the shimmering grey wall. Two minutes passed, followed by two more. *Everything's okay. She's just enjoying herself.* However, try as he might, Dr. Monier could not ease the wariness growing in him. Unable to hold still any longer, he ran after Amira into the void.

# CHAPTER ONE

## Boxes

**F**anar Monier pushed another full box against the white wall of his bedroom. He glanced around and sighed before picking up a framed photo from his desk. Anyone could see the resemblance between him and his mother. He missed the way her eyes seemed to dance when she looked at him and how her wavy dark hair fell past her shoulders. He carefully placed the photo into an open container by his door and said, "I'll find you." Then he closed the lid and magsealed it.

On the wall next to the door, the vidscreen switched from the weather report to the complex where his dad worked. Fanar turned up the volume.

"...has claimed responsibility for the vicious murder of Ryan Anderson, one of the engineers at Quasar Technologies found dead yesterday. This is the second attack from the Greeners against the company this week."

*Greeners? How can they do that? How can they kill someone like that?* Fanar's stomach churned at the news.

The broadcast continued, "Dr. James Monier, CEO of Quasar Technologies and inventor of the wormhole generator, would not meet with us but did provide a statement that the colonization project would proceed on schedule. Dr. Monier fell out of the public view for six months after an accident at the lab claimed the life of his wife, Dr. Amira Monier."

When an image of his mom appeared, Fanar turned off the vidscreen and slapped the remote down on the bedside table. A year after the disappearance of his mom and it still hurt. As if part of his heart had been ripped out.

The thick bracelet around his wrist vibrated. He released his grip on the remote and tapped his uplink. An image of Oswin Carter, his friend for nearly as long as he could remember, appeared a millimeter over the back of his hand, obscuring his skin. "Hey, Oswin."

He and Oswin were almost complete opposites in their appearance. Oswin had light skin, limp blond hair, and brilliant blue eyes, while Fanar had dark curly hair, brown eyes, and an olive complexion.

"I saw your mom on the vid. You okay?" Oswin asked, his eyebrows furled with worry.

Fanar shrugged. "I guess. You all packed?" "Most of it went yesterday, but I need your help." "With what?"

"My aquarium. The temperature won't stay up high enough. I've already lost three fish."

Fanar scowled. "You're not taking it, are you?"

"Heck, no. I just don't want to lose any more fish while we're gone."

"Why don't you give them away?"

"After all the work I put in?"

Fanar paused. Only certain boxes were going on the trip. He had to identify which ones to the people coming to gather them. Everything else was going into storage. "I don't know. I only have a couple hours before—"

"See? Two hours! If you ride your bike, you can be here in ten minutes."

Fanar frowned at Oswin.

"Okay, maybe for you twenty minutes. Still, that's a lot of time for you to come over and help me and still make it back."

Fanar sighed. Oswin's parents were both taxonomists as Fanar's mom had been. When he and Oswin had met at the age of four, Fanar had declared they should be friends, and they had been for the thirteen years since then. "Okay. I'll be right over."

He slapped the uplink to end the conversation and grabbed a bag of things he would need before taking his bicycle out of the apartment and down the stairs to the crowded streets of Fort Knox.

Formerly known to house the gold reserve for the United States, the empty vault had been repurposed to house a research facility for the newly discovered technique of artificially generating a wormhole. Stable rocky strata surrounded the cavernous underground area nearly the size of an athletic stadium, made it the ideal location to form the wormholes. When Dr. Monier and other scientists had convinced the U.N. to fund the project, work had begun. The region quickly become a modern metropolis replete with tall buildings and cramped living quarters as scientists, support staff, and other businesses moved into the area.

The street bustled with bikers and the occasional street car or autotaxi under the cloudless blue sky. As he pedaled, Fanar reflected on the news of the murder. *How could the Greeners do something so horrendous? People should not kill other people. Make the planet greener and be more responsible with the Earth's very limited resources, these were goals he could agree with... but murder?* To think he had once considered joining the organization now made him sick.

Fifteen minutes later, Fanar's hands slipped off the sweat-soaked handlebars. Breathing hard, he tapped his uplink, making a virtual display appear across the back of his hand. He tapped a couple of the virtual buttons.

Oswin's face appeared. "You on your—you look terrible. You okay?" "I'll be fine. Coming up now." He slapped the uplink and entered the building.

Oswin opened the door of his apartment, and his eyes went wide when he saw Fanar's red and sweat-dripped face. "That was fast. You sure you're okay?"

Fanar nodded as he strode into the nearly empty apartment. He'd probably pushed himself too hard, but he needed to be home in time to meet the movers. After leaning the bike up against the wall, he followed Oswin to his bedroom and collapsed onto the bed. Nearly everything else had been removed.

"Want me to get you some water?" Oswin asked.

Fanar held up a finger and took in a few deep breaths. "I'm good."

Oswin pointed to the large fish tank. "See? Only twenty-two degrees. It should be twenty-five. It was fine last night, but now it's too cold."

Six colorful fish hovered slightly below the surface, barely moving. Fanar lightly knocked on the glass, and the one nearest him moved slowly away. They'd been more lively last time he'd seen them.

He'd helped Oswin set the tank up over a year ago. The pump box connected to it had the water pump, along with multiple sensors to monitor pH, temperature, bacteria and algae counts among other things, but the aquarium they'd built had been smaller than this one. "You got a bigger tank?"

"Yeah, a few days ago. Been saving up my water rations to fill it. Mom and Dad helped too."

"That's a lot of water. What was wrong with the smaller tank?"

"You sound like a Greener. I needed the bigger tank, so I used my own water rations to fill it. I'm planning ahead since I also need the practice for my responsibilities on Niton. What's wrong with that?"

Fanar didn't respond. He knew Oswin didn't mean anything by calling him a Greener. Still, it hurt to be identified with the group that had become combative.

"So what do you think the problem is?" Oswin asked. He ran his hand over his limp blond hair. "It's the pump, isn't it?"

Fanar chuckled. "Seriously? The pump? How are you going to manage those tanks on Niton?" His friend might be more athletic and even brilliant with fish anatomy and chemistry, but his grasp on the basics of how technology worked was grossly lacking.

"I'll have you to help me," Oswin said with a smile.

Fanar toggled the power switch and closed the valves so he could work. "I don't know why you're bothering with this now."

Oswin watched as Fanar opened the housing of the box next to the tank. "After all my work, I can't just let them die. Carolina's going to take care of them while we're gone."

Inside the housing sat the water pump and another watertight box holding the array of sensors and heating element. Next to these stood a circuit board Fanar had cobbled together to make it all work. He lifted the top of the watertight box, revealing the heating element and sensors. Not seeing any visual damage, he pulled the thermistor and connected a display.

Watching the readout, he wrapped his hand around the probe. "See how the display isn't changing? We need a new thermistor."

Oswin scowled. "Can we get one today?"

"Let's find out." Fanar tapped on his uplink and virtual keys for several seconds before nodding. "Yes, they have some."

"Same place as before, right?" After seeing Fanar nod again, Oswin headed out of the room. "Great! Let's go, before I lose any more fish."

They rode their bikes to an electronics store across town. Even though Oswin took his time, Fanar struggled to keep pace. Through gasping breaths, he said, "This is why I prefer walking."

"I keep telling you to ride more. If you do, it will be easier," Oswin said. They racked their bikes at the front of the store and entered. The cramped confines had parts salvaged from otherwise useless items. The place only had five aisles, but each shelf had dozens of small containers with items tossed into them. It took nearly twenty minutes before Fanar found a box of interest. Unfortunately, the item he picked up had a crack in the casing, and the handful of others in the container weren't any better. "Will it work?" Oswin asked. "It's broken."

"With the government taking up all the good stuff, it's probably the best we can get." Fanar studied the item in his hand. "Do you have any glue left over from the old fish tank?"

"Yeah, a little."

Back at Oswin's apartment, Fanar dabbed the glue over the cracked area of the case and waited for it to set. After he was satisfied with the seal, he attached the display and dropped the probe into the water. The temperature display dropped to twenty-one point eight degrees Celsius. He removed the leads, put the thermistor in its proper place, and sealed the boxes before setting the system back into operational mode.

They watched as the water circulated through the tank. It took several minutes before the temperature readout went up. Oswin let out a whoop before slapping Fanar on the back. "Stormin!"

Fanar smiled and looked at his wrist for the time. "I'm glad it worked. I gotta go."

Back at his apartment, Fanar paused at the threshold. His entire life stood before him, boxed and labeled. Even stacked with boxes, the place felt empty without his mom's presence. Her encouraging spirit that kept him grounded was gone.

His mom had worked at Quasar along with his father. When initial aerials and tests had showed no danger, they'd pushed for the company and interested governments to allow families to travel together. She had been so excited when it got approved. Now she couldn't go. Her disappearance had caused the project to be halted for six months as surge filters and other safety measures had been put in place before another person went through. Then a few more. Fanar's mother was the only casualty.

Fanar's relationship with his father had been strained before the accident. The past ten years of his father's continued lack of presence made it difficult to have a connection. Now they barely spoke. With that kind of relationship, he had only one reason to travel to Niton. To find his mom. Logically, he knew it wasn't possible, and yet a part of him believed she was there, on Niton, waiting to be found. It didn't matter what everyone said. He knew, deep in his soul, his mother lived, and he needed to be the one to find her.

A knock on the door grabbed Fanar's attention. The vidscreen showed a man in a grey and brown camouflage military uniform. "I've come to escort you to Quasar," the man said.

Fanar opened the door, and immediately the soldier shoved Fanar aside and into the wall. The soldier rushed in and went around the apartment, his hand on his sidearm.

Fanar's heart pounded. Maybe this wasn't really a soldier, but a Greener. He rubbed the back of his head and was wondering if he should run when the man came back to the main room. Fanar was surprised to see someone in his early twenties facing him.

"Space is clear. I'm Randy. For your safety, I've come to escort you to your training."

# CHAPTER TWO

## Departure Day

*This is Fanar. Everyone traveling to Niton is supposed to create daily audible log entries, so here's mine. We're leaving today. (System mark: Departure day)*

*This is Fanar. Again. Dad said my entry wasn't adequate. Said we need everyone's experiences and feelings so future generations will know what happens. Here's what's happening. We've been cooped up in this facility for the past three months while we train over and over how to quickly set up the buildings and equipment and wait for the big day.*

*For the past ten years, they've been testing and fine-tuning the ability to create a wormhole to Earth-like planets. Oswin and I like to call it the portal. This is the first time a group of people will leave Earth to stay on Niton, the first of the nine planets selected to be colonized. There, we can get fresh water and grow food. It will be another month before they can reopen the portal to send through the return machine and additional supplies. The return machine is so we can routinely open a wormhole to*

*Earth for travel and provide fresh water and other resources. (System mark: Departure day)*

The wall's color switched from white to red, an indication the portal would open and a warning to stand back from the wall. The departure chamber grew quiet as everyone faced the front and conversations came to a stop. Fanar craned his neck to peer past the person ahead of him.

"Here we go!" said Oswin from his right. His newly grown beard and mustache moved as he smiled.

The anticipation was almost infectious. Almost. Fanar scanned the faces around him. If there was a time for the Greeners to strike, this would be it. Security stood at its highest since the project had started; however, Fanar couldn't shake the feeling that something was going to happen.

Over three hundred people and several massive transports packed the underground chamber. Fanar wrinkled his nose as the putrid odor of sweaty bodies mixed with the hot air swept past him. Despite the air system running at full capacity, the foul tang in the air couldn't be avoided. Fanar glanced up at the transport to his left. The tire, one of ten, stood at nearly double his height. Each transport silently waited. They were laden with equipment for experiments, portable housing units, common buildings, and other stuff he knew more about than he cared to. Because of his knack for technology, he had received extra training on the maintenance and repair of the items being taken, giving him very little personal time the past few months. He took a step away, not feeling any less intimidated by the monstrous vehicle.

He returned his attention to the red wall, where the portal would appear. After two minutes, a low-pitched whine filled the chamber, growing in intensity. He rubbed the hairs on his arm down as his body tingled with energy. *I must be more excited about this than I realized.*

He covered his ears and squinted. Soon everyone had their hands over their ears. The ten seconds seemed an eternity, and then suddenly he heard a pop and all went quiet again. Everyone lowered their hands. The wall became white, and those around him groaned in disappointment.

"What happened?" Oswin asked.

Fanar shrugged. His dad, having been at the head of the procession, headed across the vast chamber toward the wide door separating the departure room from the equipment powering the portal. Behind him were two men in white polyester garb, one familiar to Fanar and the other someone he'd never seen before, and a man who followed at a more sedate pace, whom Fanar recognized as Kellach, an expert with the equipment. Fanar and Oswin would be under his charge when assembling the buildings on Niton. His shoulders were hunched forward, and his thinning blond hair bounced as he walked with a bobbing gait. His clothing was expensive: rarely seen denim jeans and a natural-fiber shirt. Kellach didn't appear worried, so it couldn't be bad.

"Please stand by; we will be opening the portal shortly." The announcement boomed through the chamber, pausing conversation. After the speakers went silent, the buzz in the room resumed louder than before.

"I'm sure they'll figure it out," Fanar said. He kept glancing to the front as he talked with Oswin. He didn't see anyone exit the equipment room; instead the wall turned red, then white, and then red again.

"Is it…?" Oswin stared at the wall and covered his ears as the low-pitched noise pierced the air again.

The whine grew louder and higher in pitch then suddenly stopped, and the wall switched to white.

Fanar stared at the door his father had disappeared through. "I wonder what the problem is." *Greeners?*

"I'm sure your dad will figure it out." Oswin spun around, hitting Fanar with his backpack, and gestured to be followed. "You gotta meet this new family."

Fanar pushed the pack away. "Why?"

"You'll see. C'mon." Oswin squeezed through the crowd. "I think I saw them toward the back. The girls are our age. It will be nice having someone we can spend time with."

"I can always count on you to find girls. They're our age?" Fanar shifted his own backpack and struggled to keep up.

"Yeah, Rhona is sixteen and her sister is twenty."

"Why would we get new people when we're about to leave? They'll have missed all the fun training."

"Be nice. I don't want you scaring them away." Oswin approached a brown-skinned girl. "Rhona?"

The girl turned around, causing her curly shoulder-length hair to spin around and fly in her face. "Oswin!" She pushed her hair back and gave him a brief hug. "I'm so glad I can see you before we go through! I'm so excited I could just burst!" Her eyes went wide. "Oh my goodness, just imagine all the new things we're going to see!" She glanced at Fanar and then looked at him again. He often got second glances. He'd inherited the best of his parents' diverse heritage. His mom's flawless Saudi complexion and his dad's French build. Her smile faded a little before she grinned again. "You Fanar?"

This girl was possibly more excited about the trip than Oswin. Given the opportunity, these two would probably lead everyone. He drew the corner of his mouth up into a half smile. "I am."

"Oswin has said so much about you. Oh! This is my sister, Contessa, and my parents." She gestured to the rest of her family.

Rhona's parents introduced themselves as Terrell and LaKeisha King.

Fanar nodded to Contessa, who returned a slim smile. She stood at the same height as her sister but had a thinner figure. Her hair was straight and cut short at the nape of her neck.

Oswin snorted. "Fanar isn't as excited as we are about this. Seems—"
"Please stand by; we will be opening the portal shortly," the speakers boomed again.

"I certainly hope so." Fanar directed his comment toward the far wall, where they would disembark. "This place is starting to wear me down."

Oswin slapped him on the shoulder. "I feel your pain." He smiled at the two girls. "See, Fanar and I have been living here for several months. You're lucky just getting here a few days ago."

"Here we go." Contessa nodded her head toward the far wall. It had shifted to red again.

Fanar didn't wait for the noise this time and immediately covered his ears. The whine and the tingling sensation soon began and continued for several seconds—far longer than the first time they'd heard it.

"This is it," Fanar yelled. "Any second now it should go all grey."

As predicted, the wall went from red to grey. The whine stopped, and Fanar took down his hands to study the open portal where the wall had been. In training, he heard about it. Before today, only a few people had actually seen it, and even fewer had traveled through it. Sparkles of color rippled across the swirling grey surface with tiny flashes of white.

"Finally!" His voice sounded flat and distant. "Whoa!" Oswin grinned. "That is so weird." "Okay, everyone—"

Fanar strained to hear the announcer, who continued a few seconds later, easily heard but flat.

"Okay, everyone, the day and moment have finally come. After proceeding through the portal, remember to move off to the side. We have ample time. The portal can stay open for up to four hours, so there's no need to rush. Our first team—"

*BOOM!* An explosion knocked Fanar senseless. With his ears ringing, he shook his head and looked around. The wall was white again with some burn marks along the base.

The first thought Fanar had was that the Greeners had sabotaged the equipment. Then he remembered seeing his dad enter that room. "Dad!" He took a step forward and paused as dizziness overtook him. After a couple of seconds, he lurched toward the front.

"Fanar, wait up." Oswin opened his mouth, closed it, then opened it again. "Hold on a second. I'm all dizzy."

Fanar himself could hardly stand straight. As an afterthought, he looked at the girls. "You guys okay?"

"Yeah." Contessa faced Rhona. "You okay, Sis?"

Rhona stood still, gazing at her sister with her eyes wide. "What?" Contessa leaned in and spoke into her ear. "I said, 'you okay, Sis?'"

"No! My ears are ringing and I think I'm about to fall over." She grabbed at Contessa's arm. Contessa held her sister, barely keeping them both upright.

Fanar stumbled toward the front of the chamber, each step getting more stable. He had to reach his dad. Several people he passed were on their uplinks, attempting to get answers. Some sat, trying to recover. Oswin and the girls were on his heels. Most people they passed could be heard complaining about tinnitus or dizziness. At one point, they stopped to take a different route as someone had gotten sick.

They were halfway to their destination when they heard another announcement. "Ladies and gentlemen, our departure has been postponed. We'll leave the equipment in place. If you need medical attention, the clinic has been advised of our situation. Otherwise, check the uplinks for more information in the morning."

It wasn't a fast exit, but there was enough activity around Fanar to slow his progress. He squirmed his way through narrow gaps between the people—leaving Oswin, Contessa, and Rhona behind. Eventually, he made it to the same wide door the technicians and his dad had entered earlier. Breathing hard, Fanar found the door ajar with a thick cloud of smoke pouring out. He pushed with his hand, but the door wouldn't move.

He turned to see Oswin and the girls approaching and gestured for Oswin to help. They pushed together and opened the door. Fanar's throat and lungs burned as he inhaled the acrid smoke. He and Oswin both coughed and backed up.

"Is that what I think it is?" Contessa put her hand to her mouth and stepped back, staring at the floor past the opening. Blood had pooled just inside the doorway and had smeared across the floor to a man in a white jacket lying in the path of the door.

Fanar inwardly shrank at the sight. He believed the Greeners were responsible for another life lost.

"We'll get some medical help. Go find your dad," Oswin said.

The smoke thinned as fresh air poured in, giving Fanar a better view. Stepping over the arm of the technician, he entered the room. The space, nearly twelve meters back, took the entire length behind the wall. He pulled his shirt over his mouth. The fabric kept him from inhaling too much smoke and coughing. He didn't see a fire, but the tube along the base of the wall had a long, jagged opening across the top with parts of the inside glowing red and a liquid bubbling below the hot coils giving off wisps of smoke. A similar tube ran along the top of the wall, undamaged. A movement caught his eye, and he stepped back. *Maybe I shouldn't be in here*. He hid behind a piece of equipment and waited. He could neither hear or see anything. He stepped out.

"Dad?"

*Fsssssss!* He jumped, his heart in his throat. A billow of smoke rose from where one of the glowing coils had touched the liquid. Fanar sucked in a deep breath and immediately regretted it. He coughed again and moved away from the fresh plume of smoke. Where had his dad gone? He called out, "Dad? You okay?"

Equally spaced and perpendicular to the wall, long cabinets dominated the room. These manipulated the negative energy and used the plasma conduits to create the portal. Space near the back wall allowed him to move quickly. When he got to the other side without seeing anyone, he frowned. No door. He had come in through the only entrance.

Smoke swirled overhead as the air circulated. His heart pumping hard, he traveled along the damaged conduit tube toward the entrance, half expecting to see the charred remains of his dad lying next to it. He called out again, "Dad?" Clearly, his father wasn't in the area. Where did he go? Fear warred with hope inside him.

He glimpsed the ruined conduit. The red glow had gone. Curious, he held his hand over it. No heat.

"Don't touch it!" his dad yelled.

Fanar gasped and jerked his hand back, elated to see his father. "Dad! Where were you?"

"In the departure chamber helping everyone exit. Let's get you out into the open." He put his hand on Fanar's back, urging him into the departure chamber.

Fanar watched the body disappear as his father closed the door behind himself.

"You okay? What were you doing in there?"

"I was looking for you! What happened? When did you get out?"

"I wasn't back there! Your training should have told you it's dangerous to be back there when the wormhole is opened." He shook his head. The bags under his eyes were very pronounced and his face gaunt. The non-stop work was taking its toll, and he looked exhausted. The scowl that seemed to be ever present on his forehead deepened. "I need to get this fixed. You okay?"

"Yeah."

"Good. You and Oswin get out of here while I work."

Fanar looked around and found Oswin approaching with some medical personnel pulling a gurney and stepped aside to let the medical team follow his dad back into the room. When he got near the exit, he looked back to see the dead technician being lifted.

Kellach, the man with his father earlier, bumped Fanar's shoulder as he went past, knocking Fanar off-balance. "Sorry, bud." His voice grated.

"Hey. You okay?" Oswin asked.

Fanar turned to see Oswin staring at him. "I guess." *No, I'm not okay!* For a few terrifying moments, he'd thought he had lost his dad so soon after losing his mom. He knew the Greeners were somehow responsible.

Oswin nodded toward the dead person. "Who was it? I didn't see."

"I don't know his name. A new technician."

"Glad your dad's okay."

"He came out earlier, and I guess I missed it. You know nobody's supposed to be back there when they open the portal. It's too dangerous." Oswin scowled. "Then why was the technician back there?"

# CHAPTER THREE

## New Friends

The next morning, Fanar kept up with Oswin as they sped down the hallway. "You're in a rush to visit these girls, Oswin."

"Rush? I don't think so. Besides, you want to make a good impression so we have some friends to pass the time with."

"Which habitat did Rhona say they're in?" Fanar stopped, happy for the respite.

"Five?"

Fanar rolled his eyes. "Your confidence is overwhelming. Oh, wait… your confidence isn't 'adequate.'"

Oswin chuckled and tapped his uplink. "Where is Rhona King staying?"

A male voice spoke from the bracelet. "Rhona is registered to habitat five. Would you like to be guided?"

"Yes."

"Follow the green circle."

Under Oswin's feet, a bright green circle appeared in the dark grey material of the floor and flashed two times before it moved forward a meter and stopped. As they proceeded through the light-grey-walled hallways, the circle moved to maintain a constant distance in front of them.

When they were twenty-five meters away from their destination, Oswin's uplink vibrated. He glanced at it and tapped a button to indicate he wanted Rhona to be notified of his arrival. At the habitat, the circle blinked and disappeared.

The door they approached opened, and Contessa poked her head out. "Oswin! Fanar! Come in, have a seat." She led them inside and sat on the couch next to Rhona.

Mrs. King stood next to Rhona, her eyebrows furled with worry. "I'm going to check on your father." She gave a brief wave at the boys and put her hand on Contessa's shoulder before leaving.

Rhona watched her mom leave then focused on Fanar. Her eyes, so alive the previous day, were sullen. "Think we'll be able to leave tomorrow?"

"I don't know. I hope so," Fanar said as he sat across from her. "I only saw one item damaged."

"I had hoped you would have some inside information."

He shook his head. "Sorry. My dad doesn't spend much time… explaining things to me." He swallowed and put his foot up onto his opposite knee. "I'm sure it'll be fine. Your mom okay? She looked upset." "We all are. Daddy's been in the infirmary. If he doesn't get well, we won't go."

Fanar frowned. "I'm surprised you're here at all. How did you become a part of this?"

"Our parents were hired at the last minute," Contessa replied. "One day we're talking about a vacation, and then a few days later Daddy says we need to store everything because we're going to the new world."

"Oh, my goodness"—Rhona sat up—"do you know how hard it is to figure out what you're going to take on a two-year trip? And with such limited space."

Oswin nodded. "That can be hard."

"Hard? Try impossible. Clothes and shoes took up almost all my allotted space."

"Hair products." Contessa shook her head. "Are they going to manufacture that?"

Fanar grinned. "We'll have the portal opened at regular intervals for fresh supplies."

"'Portal.' Everyone keeps calling it the portal. Isn't it a wormhole?" Contessa asked.

Fanar glanced at Oswin, who returned his grin.

Oswin motioned with his finger between himself and Fanar. "I'm not sure who thought of it, but one of us did. We thought it would be fun to give it a new name. Soon others started to call it the portal, and, well, I guess it stuck."

"Everyone except my dad," Fanar replied.

"Yeah. Not him," Oswin agreed.

Rhona grinned. "I guess you get to pick out the cool names for things. What about you? You pack anything special?"

Fanar spoke in a low voice. "A picture. A picture of my mom."

"Ooooh, a picture." Rhona glanced down at her uplink. "I think I have a few thousand of those."

Contessa gripped Rhona's arm. "I think he means on paper."

Rhona's eyes went wide. "On paper?"

Fanar nodded.

"That must have cost a fortune."

Fanar nodded again. "I had to save up for two months. I wanted something more... solid."

Contessa let go of Rhona's arm and nodded as Fanar spoke. "Yeah. Something with real value."

"Yeah. That's it exactly."

"I packed a porcelain doll," Contessa said. "It's been in my family for six generations. My grandmother gave it to me. She said it was one of the few remaining. I couldn't just put it into storage."

Rhona grinned. "I packed chocolate."

Contessa gasped. "Chocolate?"

"Gotta have my chocolate."

Fanar laughed. "I wonder how long it will be before others start asking for some."

Rhona folded her arms. "They better not." She looked at Oswin. "What about you?"

He shrugged. "Some clothes. Couldn't fit my aquarium."

"Wow. I bet you're a cheap date."

"Well, I don't know about that."

"What do you do for fun?"

"Camping and working on my aquarium. Guess I follow in my parents' footsteps. They're taxonomists, and I like to study fish."

Rhona's mouth hung open.

"Well, someone has to. What do your parents do?"

Rhona held her head high, obviously proud of her family. "Mom is a data archivist and Daddy is a munitions expert. Best in the field."

Fanar straightened up and set both feet on the floor. "Explosives?"

"Yeah."

Fanar narrowed his eyes. "So he's good at blowing things up?"

Contessa met his gaze. "What are you saying?"

Rhona sat forward and raised her voice. "Yeah. What are you saying?" She put her right hand on her hip with her fingers facing backward. "You think he had something to do with the explosion yesterday? He was with us. Remember?"

Oswin started, "Fanar—"

Fanar interrupted him. "Bombs can be set." He couldn't stop. The anxiety of the past few days and his fear of the Greeners poured out. "Lots of people are out there protesting against this."

Contessa drew her mouth into a thin line, breathing heavily through her nose. She measured her words out. "He got hired to help on Niton. They said some minerals need to be mined. He's an expert."

"An expert with explosives."

"I think you should go."

Rhona stood. "Yes, but first he needs to apologize."

Fanar pulled on Oswin's arm as he stood. "Let's go, Oswin. I think we need to talk to my dad about this." He turned to leave.

"We could have been killed!" Rhona yelled at his back.

"What is your problem?" Oswin asked as they left. "I don't think they're terrorists."

Fanar lowered his voice. "Not terrorists. Greeners. Or at least I think their dad is. We know they don't want us to succeed. Don't want us to 'rape other worlds' as they say. And now they might not go because he's sick? I was back there after the explosion. Only one thing was destroyed. Not enough to kill us; just enough to keep us from leaving. Think about it." Fanar took quick long strides. "Did we hear anything about mining in our training? No. Just water and arable land. How many times did we open the portal this past year without problems?"

"Let's see—"

"Twelve times. Once per month because it takes a month to get the power reserves without upsetting WEC. Twelve times they sent drones to Niton, and we didn't have any issues. An explosives expert is here for a few days"—he glanced at Oswin—"and we suddenly have a problem. Remember the new guy caught in the explosion? I bet he stayed behind to set the bomb."

"Maybe you're right; I don't know. You definitely need to talk to your dad about this and see what he thinks."

Fanar passed through the departure chamber, giving space between himself and the large transports that stood as giant reminders of the failed attempt to leave. Soldiers stood guard at the engineering room, allowing some people in after they showed badges. His dad stood outside the room with another engineer, facing two men in dark suits.

As Fanar approached, one of the suited men said, "If the U.S. can't handle this project, then maybe we should move it to a country that can!" He and his companion departed, their strides long and swift. Dr. Monier's face and shoulders sagged while watching the retreating figures.

"Fanar?" Kellach came up from behind. "James' son, right?" He wiped his thick hand on his pants and shook Fanar's hand. "Sorry about earlier. Not every day you learn you've lost your new right-hand man. Right?" He pushed his thinning blond hair away from his round face.

"I guess not."

Kellach looked at the nearby door. "Me? I get to replace the damaged equipment." Slouching, he walked to the door and showed his credentials to the soldier before walking in.

Fanar's dad addressed the technician next to him and snapped, "The low level isn't adequate. Run the signal cycle again at one percent. Each station!"

Fanar winced. It wasn't like his father to yell at someone. He must be under intense pressure to make this all work.

"That's a lot of energy. WEC won't like it." The technician noticed Fanar and inclined his head in acknowledgment.

"Let me deal with the World Energy Council. I have to talk to them today about getting a redirection of power so we can do this in the next couple of days."

The technician snorted.

"They report to people who want this project to succeed. They won't like it, but they'll do it. If I'm right, then only the conduit needs to be replaced, and we can move out." Fanar's dad turned. "Oh. Fanar, I'm really busy—"

"This is important," Fanar said.

"We need this today, Reynolds," his father said to the technician and came to Fanar. "What's on your mind? You need to make it fast."

"It's about the explosion. Do you know a man named Terrell King?"

"Terrell King? Yes, he's someone I hired on last week. Why?"

"You hired him?"

"Personally. Very important he comes to explore the possibility of an immediate mining operation."

Fanar sputtered. "Mining? We never talked about mining."

"There was always the probability of mining; we just thought it would be significantly later, after a detailed survey. Our situation on Earth is more desperate than your average citizen realizes. If you would stop tinkering and pay attention to what's important, you would know we selected this planet as the first to colonize after its discovery with the deep-space telescope because it showed not just an abundant supply of water, but the promise of rare resources. We found an important element in the last sample drone, so we decided to bring an expert. This family will need some extra support since they skipped the training. Can I depend on you to help?"

Fanar watched Kellach exit the engineering room with a portion of the damaged conduit. He looked at his dad, sighed, and rubbed his forehead. "Yeah."

"Good. Always remember, you do the right thing…"

Fanar finished with his dad, "And everything will turn out right."

"Exactly. The plasma conduit fatigued faster than calculated, and that's what caused the explosion. We have a spare we need to install. I'll see you soon."

Fanar watched his dad walk away and approach Kellach. He felt like he had been punched in the gut. Do the right thing? He had just accused Mr. King of being a Greener. How could he face Contessa and Rhona after his crazy accusation?

# CHAPTER FOUR

## Landing

*This is Fanar. It's been two days since the explosion. We are leaving first thing tomorrow morning. I must admit, I'm a little nervous. What if there's an explosion while we're passing through the portal? Oswin hasn't said two words to me since I told him what my dad said. Making it worse, Mr. King is fine and traveling to Niton. I was hoping I wouldn't have to face Contessa and Rhona again. (System mark: Departure Recovery +2)*

Fanar stared at the ceiling from the comfortable position of the top bunk. He and Oswin were in a bunk room with several others. If he closed his eyes, maybe he could ignore the rustling of people getting ready for departure.

"Fanar, let's go," Oswin said to him as he climbed out from his bottom bunk. Okay, maybe Fanar couldn't ignore everyone after all. At least Oswin was talking to him.

Fanar climbed down and pulled his shirt and pants from the drawer. All his clothing had been stored on a transport except for a spare change of clothing in his pack. Because personal items wouldn't be available the first few days on Niton, he really wanted to keep those items clean. He sniffed his shirt and crinkled his nose. It wasn't going to be pleasant in the cafeteria. Of course, he should be used to it. To conserve water, he could only shower up to three times a week. Some areas of the world were limited to once a week.

Fanar glanced at Oswin and found him staring back. Oswin didn't say anything as he turned and walked away. Fanar sighed and followed. Oswin was still unhappy despite Fanar's sincere apology to the girls. It didn't do any good though. They weren't talking to him either.

They made their way to the cafeteria for breakfast, where his prediction proved true. The pungent cloud of body odor thinly masked by deodorant permeated the space. However, as time passed, he didn't notice it anymore. A stack of breakfast bars next to bottles of water waited for them. No fruit today. He took his ration and sat next to Oswin.

He tried to smooth things over a little with Oswin. "Maybe Greeners didn't cause the explosion. Maybe it's like my dad said. The tubing had decayed from all the energy hitting it, and the timing was a coincidence." Fanar took a bite of his bar.

Oswin stared past Fanar.

Fanar knew what Oswin stared at, or rather whom.

"You know…" Oswin spoke without looking at Fanar. "I shouldn't be mad. After all, I half believed you. It made sense. But did you need to accuse them to their faces?" He faced Fanar.

Fanar turned away and swallowed, feeling the lump scratch all the way down. When it finally reached his stomach, he became nauseated. The water didn't help. "I'm going down to the departure chamber. I understand if you don't want to be with me." He stood and left. On the way out, he dropped the remaining portion of his breakfast bar in the bin for organic waste. He would have dropped the water in too, but it was too precious. He had traveled several meters down the hall when he heard Oswin call his name from the cafeteria exit.

Oswin approached with his head down until he came up next to Fanar.

Fanar held his gaze.

"I'm done eating."

Fanar nodded and continued toward the departure chamber. He took a deep breath, and the knot in his stomach relaxed.

A brief warning poured out of the speakers before the wall shifted to red. After extensive testing the prior evening by the engineers, the portal was ready. Fanar wasn't sure how they'd done it, but his dad had come through and gotten the surge of energy needed to open the portal. It wouldn't surprise him if it involved rolling blackouts across several cities, something that would  magnify the antagonistic attitude toward the project. Not everyone believed this endeavor would benefit mankind. Fanar covered his ears, doing his best to ignore the tingling all over his body. Clearly it had something to do with the portal.

Once the portal became active, the loudspeakers boomed, "Ladies and gentlemen. Another reminder to move at a moderate pace through the portal. When you get to the other side, don't stop. Move forward and to the side so others may come through without mishap." The speaker paused. "Team one, you are free to proceed."

The first team consisted of six soldiers known as the 'security team.' Fanar knew his dad hated it, but their presence was non-negotiable. They wore full battle gear and carried rifles and looked, to Fanar, rather silly. Only they knew why they needed to be battle ready, for the drones showed only plants and small animals. He recognized Randy, the one who had escorted him from his home, in the group.

"There they go." Oswin spoke into Fanar's ear so he could be heard.

"They are going to be really bored."

Fanar grinned, really glad Oswin was talking to him again. "They'll help with setup."

"But what then?"

Fanar observed the expectant faces around him then back to the shimmering greyness. So far, his dad's theory held. No explosion or any other indication of a problem.

The security team neared the wall, and the leader paused before putting a foot forward. Like moving out of sight behind a curtain on stage, he disappeared into the void. The other soldiers were right on his heels. Everyone waited. After several seconds, the leader came back into view. He smiled and waved.

Fanar released his breath and cheered along with everyone else. He didn't even realize he'd been holding it. He kept cheering as the soldier spun around and disappeared through the portal. After the security team, Fanar's father and a few others stepped through. Fanar scoffed to himself. *Do the right thing? What about being with your family?* He focused on the exit. If he left, would his dad notice? Still, he had to see what was on the other side. Would he see what his mom had seen? Even now, he had doubts of finding her, but he had to try.

Following the initial group of people, each encumbered transport approached at a turtle's pace and disappeared through the opening, followed by an assigned group of people. After watching a long while, Fanar glanced at the countdown timer prominently displayed above the exit. Only thirty minutes left.

"Relax, bud. There's only one more platform." It was Kellach, who had just joined them. "Where's your dad?"

"He was in the first group to go through after the soldiers," Fanar said. *Work first, family second.* Clearly his dad didn't need him, and he didn't need his dad.

Kellach nodded then asked Oswin, "How about you?" Oswin answered, "They went through an hour ago."

"Me? I'm glad to be back here," Kellach said. "Let them go first. Then if there's a problem and the portal shuts down, we won't be stuck there." He winked at them and faced the front.

"That would be bad," Fanar said. Oswin nodded in agreement.

The final platform approached the opening. Fanar walked in step with everyone as the substantial vehicle slowly moved into the shifting grey emptiness. He glanced at the timer. Fifteen minutes. Kellach was right— plenty of time. The platform disappeared. People walked through, and Fanar got closer. While the wall could fit only one platform at a time, forty people could go through at once. He glanced at Oswin and saw a grin plastered on his face as he stared ahead.

Fanar studied the shimmering wall for a half second before holding his breath and going forward.

For a brief moment, the area around him had a misty quality. A haze he was suspended in before he felt an urgent pull. The tingles became a thousand painful pinpricks on his skin. Then he stood on solid ground. The pain quickly abated like a lingering nightmare.

"…can arrive safely. Keep moving forward and to the side so others can arrive safely." The message repeated as they came through. If not for the reminder, Fanar would have definitely stopped.

The air was damp and cool, maybe nineteen degrees Celsius. But the lake made him hesitate. A few kilometers away, it dominated the view with the land gently sloping down and around it on the east. On the right, the land continued westward to hug the lake until both met mountains poking up into the horizon. He knew from the aerials that the south end of the lake ended with a waterfall and forest he and Oswin were planning to visit, while the west side of the lake met the mountain range where a couple of rivers fed into it.

"Wow." Oswin's statement echoed Fanar's thought.

Fanar's head pivoted to peer in all directions as he walked to their predetermined spot. The plum-colored grass-like vegetation covered the land as far as he could see. The vista resembled the images they'd seen in the 360° projection room, but like any image, it had lacked the personal experience. The smell in the air, the barely present breeze, the warming of his skin from the sun, the hint of unknown dangers lurking in the lake and in the mountains beyond.

Fanar spun around to look up beyond where they'd arrived. He couldn't tell how far the land went. The ground continued up gradually until it met the deep blue sky in the distance. He turned to the grey emptiness of the portal shimmering in the middle of open land. Nobody came through. Then, in the blink of an eye, the purplish ground cover remained. The portal gone. He shifted from leg to leg and squatted a few times.

Oswin stared at him. "What are you doing?"

Fanar had never thought of himself as heavy, only sixty-five kilograms, but with the stronger gravity, he could feel every kilogram of his twenty percent weight gain. It wasn't like the weight belts they'd been forced to wear in training. "Seeing how much the extra gravity is affecting me. I just gained thirteen kilos!"

Oswin swayed from left to right and back again. "Yeah, stormin'!" Kellach came up to them. "Guys, time to get busy."

Fanar took one last glance at the lake and tried to swallow the lump that formed in his throat. He could finally admit it to himself: he had hoped she would be standing here, waiting for him. They'd never known what happened to his mom. She had been the first to step through, and she'd gone by herself. She had wanted to see it without any bias from others who might travel with her. As the four-hour window had drawn to an end, his father and then others had gone through and hadn't been able to find her. As Fanar gawked at the sloping landscape, he considered the possibility she wandered off to explore and lost track of time. He reminded himself that she had been on many campouts and had survival skills. With that encouragement, he turned to do the work given to him.

His platform carried the rovers, repair equipment, spare parts, a maintenance building to house it all, and much more. Groups of people started the task of assembling other common buildings that were arranged in a large circle, the maintenance building at the northmost point.

One of the cranes mounted on the platform lifted a crated rover and set it on the ground. If all went well, he and Oswin would be taking it out in a few days for some exploring of their own. Underneath were the aluminum walls, flooring, and mounting posts of the maintenance and storage building and other items. A second crane attached to the platform had a drill rigged to it. It reached out and drilled holes into the ground and then lowered the mounting posts into each hole. Some for the center of the building and then at the four outermost corners. After some plascrete was poured around the posts, Fanar, Oswin, and other colonists held the outer mounting posts as floor panels were lowered into place and fastened. In minutes, the plascrete had set.

For a while, Fanar was nauseous. The heavier gravity affected him in ways he hadn't expected. He could only work for twenty to thirty minutes before needing a quick break. After a couple of hours, the nausea went away, but he was weak. He wiped the sweat off his forehead and stared at the lake.

"It just calls to you, doesn't it?" Kellach strode up to Fanar and Oswin. "Take a good look, boys, because this is why we're here. Clean water with no pesticides, plastic, pharmaceuticals, or heavy metals." He put a meaty hand on Fanar's shoulder and the other on Oswin's. "And it has a neutral pH too. Me? I feel it's too beautiful to touch."

He approached the platform and retrieved some water bottles from a side storage bay. After handing one to each boy, he held his up to the sun and stared at it.

Fanar looked at Oswin, who widened his eyes and shrugged. Fanar smiled and shrugged back.

Kellach lowered his bottle and held it out to the lake. "Drink up, boys, because soon we'll be drinking from there instead."

He took several swallows and shifted his gaze to the work in progress. "No! That's not right." Kellach ran over to where people were connecting walls.

"I think he's been around the high energy field a few times too many," Oswin said as Kellach ran away.

Fanar chuckled. "Probably, but I like him."

After their break and several more hours of work, Fanar collapsed onto the ground and peered up at the sunlit sky. "Think you'll get used to the longer days?"

"Twenty-nine hours, four minutes and ten seconds." Oswin groaned. "No. Well, maybe eventually. Exciting, huh?"

"I can hardly wait to get my hands on the technology we brought." Fanar rubbed his neck and looked around. Oswin had closed his eyes, enjoying the rest. Past the immediate area, another building was being erected for the computer core and access terminals, leaving space between it and the maintenance building for where the circular road would be laid. Beyond that, the wide-open space. The openness intimated him. No trees or buildings pressing in. Everyone focused on their own task or rested. Still, he couldn't shake the feeling of being watched.

A little more than fifty-five meters south of the maintenance building stood the newly erected meeting hall. Large enough for everyone, it was the giant of all the structures being assembled. Fanar saw two figures talking to each other—Contessa and Rhona. He could just barely make out who they were. As if sensing his gaze, Contessa glanced his direction then resumed talking with Rhona.

"I'm sure they'll forgive you." Oswin sat up and took a drink of water. "It just might be in a couple of years." He smiled.

# CHAPTER FIVE

## Sabotage

*Oswin Carter's log entry. I'm sure everyone else is saying the same thing about our uplinks. There's something here keeping normal radio signals from working, so they're no good unless we're standing near the computer building. It has the latest system called the IQ5000. Seriously? We'll have to give it a fun name sometime soon.*

*For the past two weeks, we've been sleeping on cots in the public buildings as we assemble the houses. It will be nice to live in a normal space with some privacy. I'm not sure when I should be awake or asleep with the longer Nitonian days. I can hardly wait to see what kind of creatures live in the lake and explore the waterfall. (System mark: Day 14)*

swin stepped back to admire the finished work. Before coming to Niton, the settlers had spent several months learning how to quickly construct the buildings, operate the special equipment, and get associated with what they should expect on the new planet. The long days of training were finally paying off with the strange building in front of him. "I still say it's not like any house I've seen."

The round two-bedroom house, topped with a domed roof, had a footprint of a little over a hundred square meters. The framework centered on a post supporting the entire structure. Like all the buildings, its roof was made of solar paneling and could divert rainwater into tanks for later use as grey water.

He stood with his parents, Doctors Garrick and Candice Carter.

"These are incredible structures," his father said and took a few steps toward the house.

Oswin had always heard how he resembled his dad and dismissed it. Now that he had matured, it was like seeing an older version of himself in the mirror. Both wore a close-cropped mustache and beard. Of course, his dad had more grey in his than light brown, though he still maintained a slim figure.

"Dr. Monier says they're a lot like… what did he call them?" his father asked.

"Um… Dixon? Dynason?" His mother searched for the right response.

Oswin's mom put her arm around his waist. There was a time he would have pulled away, but now he was happy for the closeness he had with his parents.

"Dymaxion. That's it!" his father exclaimed. "Proposed houses from way back in the mid-nineteen hundreds. They may appear odd, but these structures are designed to easily withstand high winds and are very efficient. We had discussed constructing a large building to house everyone, but it was believed for ecological reasons that small houses would be better." He verified nobody else was close and whispered, "Actually, there were a couple of board members who strongly opposed living in a dormitory, so we had these made instead."

"That's great, Dad, but let's focus on the more important thing."

"Like what?"

"Like when can we eat?"

Oswin's mom chuckled and gave him a squeeze before letting go. "Let's try it out."

She led the way into the house and opened the freshly stocked pantry to reveal several packages. "This sounds good. Chicken flavored protein patty with an apricot flavored sauce." She turned to Oswin. "Comes with vegetables. Sound good?"

"Sounds good."

*Pah.* The air rushed into the vacuum-packed container as she pulled the top off before placing it into the cooker. The appliance worked with an older technology, a combination of infrared light and microwaves to cook the food. Several buttons lined the side of the panel, but they were rarely used because prepared food containers were immediately recognized by the software and cooked appropriately.

His mom tapped the start button. Immediately sparks flew out from the panel, causing her to shriek and jump back. She stood in horror as the sparks continued to spew and flames erupted.

Oswin ran past her and opened a small door in the wall, revealing an electrical panel. He tapped all the breakers, making them switch from green to red, and turned back around in the darkened room. His dad aimed a device resembling a small cannon and pulled the trigger. The flames diminished under the assault of a pulsating sound emitting from the extinguisher until nothing but a blackened appliance remained. He grunted. "Anyone for blackened food bricks?"

The interconnected road panels glowed softly. Heavy panels about five by eight meters, twenty-five centimeters thick, connected to form a circular road around the large common buildings with several branches reaching out, lined with homes and a lab at the end of each. Like every other flat surface, the roads had the ability to convert and store solar energy. Water channels allowed water to flow through to the ground beneath them.

In the sky, only two of the three moons were visible. The larger one at half-moon, called Koch, and the smaller one at full moon, Kepler. The missing moon bore the name Hoyle. All named after astronomers, of course. Oswin walked across the grass, wet with dew, to the maintenance building. When he opened the front door, he faced a crowd of at least twenty people. He spotted Fanar and approached him. "What's going on?"

"More problems. Our inverter shorted out the second we plugged it into the solar coupling. You?"

"Our cooker caught fire when my mom tried to use it." "She okay?"

"Yeah, just scared her. Then we discovered the power surge also cooked our breaker panel and several other pieces of equipment."

Fanar laughed.

"What… oh! Yeah, the cooker got cooked." Oswin chuckled at the unintentional pun.

Kellach stood behind the countertop, entering information on repairs, promising to get to it soon. Eventually he addressed them. "Fanar and Oswin, my favorite workmen. I've been taking down information for the past three hours. Me? I could use a break. You mind holding off the crowd for a minute?"

Oswin realized he and Fanar were the last ones. "Sure." Kellach disappeared through a door behind the counter.

"So how was the food?" Fanar smiled.

"Extra crispy, I suppose. I never had a chance to try it. I heard others were having issues but never realized it was so bad. He's been at this for three hours?"

Fanar looked around and whispered, "Do you think someone in our group is a Greener? Maybe they did it."

Oswin's heart sank. *If one of the settlers is a Greener, who could we trust?* "I hope not. I can't imagine spending the next two years being afraid of everyone around me."

"I guess we can depend on the soldiers. Right? Sorry, the 'security team.'" Fanar made air quotes with his fingers.

"Ahhhh. Much better." Kellach appeared in the doorway. "Now what can I do for you fellas?"

"Cooker, breaker panel, and some other equipment."

"Power inverter," Fanar added.

"Huh." Kellach studied the virtual display. "Seems no matter where we go, we bring trouble with us." He looked up. "Gotta wonder why we bother, don't ya?" He continued to enter information.

"What do you mean?" Oswin asked.

Kellach tapped the counter, and the display went away. "Oh, don't mind me. I was just talking. I gotta be honest with you. We don't have the parts. The board will be meeting tomorrow morning to discuss options. Until then, make do the best you can."

Oswin frowned and walked out the door. *Could it be a Greener?*

Fanar followed and asked, "You still have food you can eat?"

"Yeah. We'll be okay." Oswin looked out toward the distant lake. The moons were brighter now, and their light reflected off the lake. He turned to Fanar. "Have you been down to the lake yet?"

"I haven't had time. You?"

"No. Let's go."

"Now?" Fanar asked.

Oswin gazed at the lake. Since he'd arrived, it had been calling to him. How many new species of aquatic life would he find there? "Sure. Unless you're doing something else? Besides, I'll need to get there eventually to survey the marine life. Might as well start now."

Fanar smiled. "Okay."

"We can run there."

Fanar rolled his eyes.

Oswin tapped his uplink then grimaced and slapped his hand on it, dismissing the virtual display. "I keep forgetting we can't depend on these so much. Let's tell my parents first." He ran toward his house and Fanar followed at a more sedate pace.

The whine of the water pump kept the silence at bay. The size of a rover, the water pump sat in the shallow water, utilizing an experimental process of using the water it pumped to power itself for pumping and filtering the water. Six plump hoses carried potable water back to the colony. A couple were for the colonists; the others were connected to bladders on the platforms so nineteen thousand kiloliters of fresh water could be sent back to Earth each month.

"How deep do you think the lake is?" Oswin picked up a pebble at the lake's edge and threw it out as far as he could. *Plink.* He watched the ripples reflect the moonlight as they spread and grew smaller until a few lapped against the shore.

"About fifty meters," Fanar replied. "That's what one of the drones recorded. You know, after the first one got lost. Wouldn't it be fun if we could find it when we go exploring?"

"What was it supposed to do in the event of a lost control signal? Because if it was supposed to land and wait, then it's down there at the bottom of the lake. Maybe it just kept going and it flew 'til it ran out of power. We'd never find it, then."

The boys grew quiet as they stared out at the lake. In the moonlit night, the colorful plants appeared as shades of grey. If it weren't for the two moons hanging in the sky, they could have been friends back on Earth visiting a lake.

Oswin tossed another stone. "What do you think about the equipment failing?"

"I have an idea on that." Fanar picked up a stone and cast it over the lake. *Kerplunk.* "It had to be done at least a week before we left. The guy who died in the explosion was new. What if he was a spy for the Greeners? Maybe he did it and then did something to the portal generator."

A Greener on Earth was more palatable to Oswin than the thought of one here on Niton with them. "Maybe." He chucked another stone and watched the water dance. "But why? What is gained?"

"I don't know. A demonstration. Another way of saying we shouldn't be here." Fanar sniffed. "Can we head back? I think some of the plants are rotting."

Oswin nodded and turned toward the colony. "I wondered what that smell was. Who would know what the lost drone was programmed to do?"

"I doubt I'll see my dad soon enough to ask him—" Fanar came up beside him and stopped. He tilted his ear toward the lake.

"What?"

"I thought I heard something." Fanar listened while scanning the shoreline before shaking his head. "I don't know; maybe it's just knowing I'm on a new planet. I feel like someone's always watching me."

Oswin nodded in understanding, although he didn't share the feeling. "Let's go; the smell seems to be getting worse." Fanar strode toward the settlement. "You know, I bet Kellach would have an idea about the drone."

"Yeah! Let's ask him in the morning."

Oswin met Fanar on the soft-glowing road the next day. The sun still hadn't risen; it wouldn't until later, after he was asleep. *Will I ever get used to these longer days?* "Did you see your dad? Were you able to ask about the drone?"

"No. I'm not even sure he came in last night. Here. Since your cooker is dead." He held out a breakfast bar, steam rising from it.

"Thanks." Oswin took a bite and immediately spit it out into his hand and breathed quickly through his open mouth. *That was hot!*

Fanar laughed. "You idiot. It's literally steaming."

"I know! I thought maybe it was just steaming because of the morning air." He put the piece back into his mouth and breathed through his mouth while chewing it. "Mmmm. That's good."

"When are we exploring? I've been wanting to try out one of those rovers. I hear they have some great tech."

Oswin contemplated the dark hillside. "Definitely when my sleep schedule coincides with daylight. A couple of days? That will also give me time to see some of the local aquatic life." He grinned at Fanar.

"Sounds good to me. I'm pretty sure we deserve time off after fourteen days straight of work."

"I had a thought—" Oswin paused, considering how to best present his idea. "How would you feel about inviting Contessa and Rhona along with us?"

Fanar took in a long breath and slowly let it out. "You want to do that?"

"I think it would go a long way to mending the damage."

Fanar kept walking in silence for nearly ten seconds before he responded. "Okay."

"You don't seem adequately convinced." Oswin smiled at his friend.

"No, I'm good. I'm good."

Arriving at the maintenance building, they found it empty. Fanar went around the front counter and peered through the doorway into the back. "Hello?"

"Kellach said there's a meeting today with the board. He's probably at the meeting hall."

They walked down to the meeting hall, where a few other colonists entered ahead of them. At the front table sat the board of doctors, Christine Huff, Paul Williams, Aiko Yamamoto, Anton Chernoff, and James Monier. Nearby, Kellach stood, listening to Dr. Monier address the attendees. Fanar's dad appeared as though he hadn't slept in days.

"...last eight months of testing and verifying the status and validity of the supplies to be brought over to Niton. We can overcome this. It's a challenge—no, a calling—that I feel we are adequately prepared for."

Oswin grinned at Fanar, who grinned back. Dr. Monier's favorite word seemed to creep into every conversation. It was a source of amusement between the boys. Sometimes they would listen just to see how long it took before he said 'adequate.'

"Are you saying that someone sabotaged the equipment after we arrived?"

Oswin couldn't make out who asked the question, but it caused an immediate murmur of conversation to flood the room. He froze. If the equipment failures were because of sabotage after arriving at Niton, then whom could they trust? Fanar's eyes were fixed on him. Oswin grimaced. Would Fanar be blaming Mr. King again?

"Everyone! Everyone!" Dr. Monier raised his voice to get their attention. Slowly, the noise diminished. "The last thing we need is for this to become a witch hunt. What I'm saying is that it most likely happened at one of the suppliers by someone opposed to our mission after we approved the initial test items, and we will bring this to the attention of those Earth-side when the wormhole is opened again. You have my word. Our immediate concern is to address the housing situation. Several structures were affected in a variety of ways, so we are asking for people to share accommodations where possible. We'll gut some structures to make others useful. It will be tight for a short while. We have fifteen days before Earth opens the wormhole to deliver the small wormhole generator along with other items. At that time, we'll give an update and arrange for another opening of the portal before they move on to the next planet. Thank you."

*Shared housing.* Oswin sighed. *So much for my private space.* Dr. Monier meandered through the crowd, stopping to talk to and reassure various individuals. He was the leader. Everyone counted on him for guidance and safety. Sure, the soldiers were available, but he was the one everyone leaned on for the certainty of a successful mission.

"No, I'm not exempt." He had stopped again to talk to someone nearby. "Authority doesn't give me privilege. If anything, it just gives me more responsibility and the need to lead by example." He pointed at Oswin. "We'll be taking on the Carters."

# CHAPTER SIX

## The Falls

*Log entry for Contessa. I think I'm getting used to the longer days. We'll have a couple days of darkness while we're awake. When we wake up the next day, it will be bright out and the natural day will be half over. Then we can enjoy about two days of time where daylight almost matches our wakefulness. After that, we go to sleep while it's still light out before the cycle starts again.*

*A few days ago, Oswin and Fanar invited Rhona and me to go exploring. We're going tomorrow when our day coincides with the natural daylight. Rhona's very excited and willing to give Fanar a second chance. Personally, I forgave him long ago. Rhona seems to think she's going to discover everything about the new planet herself. I'm curious but not sure if I should leave Mama. She needs me. (System mark: Day 23)*

**C**ontessa woke to Rhona snoring. *For someone wanting to be an explorer, she sure likes to sleep.* Contessa peeled back her bedcover and sat up, putting her feet onto the floor. The ceiling automatically glowed in response to her movement. It was just enough to keep an occupant from running into something. She covered her mouth as she yawned and stood. She could ask for brighter lights but didn't want to disturb her sister. She silently slid the door open halfway, squeezed through, and closed it. It was already bright outside; hopefully it didn't wake Rhona. They were supposed to get an early start, but Contessa could use some time to herself.

She glided through the hallway that wrapped the inside of the quiet house, past the closed door to her parents' room and the dining area, arriving at the kitchen. An empty glass sat on the counter. Probably left by her dad. He never put his dirty dishes away. This meant he was already up and out to do whatever Dr. Monier had in mind for him. Probably investigating the site her dad had mentioned yesterday as a possibility for mining some rare minerals. The man they'd taken in during the house consolidation was also missing from the couch. He and her father would probably be gone for a few days.

Contessa put the glass into the dishwasher before getting a clean one out of the cabinet. She held it under the water dispenser and watched it pour water into the glass until it reached two-thirds full and it automatically stopped. She let it sit there as she gazed at it. They'd run out of the water from Earth yesterday afternoon. They claimed the Nitonian water, filtered for contaminates and microorganisms, to be safe. She was suspicious but really thirsty.

With a sigh, she picked up the glass and took a sip. It tasted okay. She took a swallow. Not only did it taste okay, it actually tasted good. She finished the glass and put it under the dispenser again. After it filled, she hastily picked it up and drank half its contents. She contemplated the remaining liquid. *Best to not waste it.* She finished the water and closed her eyes. She couldn't remember the last time she had consumed so much water at once.

"Good morning."

Contessa jerked at the sound of her mother's voice and spun around. *Did she see me drink two glasses?* "Mama."

Her mother looked at the glass in Contessa's hand. "So? What's it like?" Contessa glanced down at the glass and then back to her mother.

"Good. Tastier than I thought it would be. How are you? You want some water?"

She reached toward the cabinet.

"No, I'm good." Her mother smiled, deep lines near her eyes crinkling. She appeared well. Rested. An unusual thing with her chronic fatigue syndrome.

"Not sore? Because I can stay if you need me to."

Her mom put her hand on Contessa's cheek. "No. You go and have fun. I'll be fine." She dropped her hand. "You know that school they have for the younger kids?"

Contessa nodded.

"I'll be spending some time today with Julie Shaw, the teacher. Why don't you wake up your sister? I'm sure the boys are anxious to go."

"Okay." Contessa put her empty glass into the dishwasher and kissed her mom's cheek before leaving the kitchen.

Rhona still slept. Contessa left the door open and made her bed before turning to her sister. Rhona had covered her head against the sunlight pouring in through the doorway. "Rhona. Time to get up." She waited two seconds for a response. "Okay, I'll go with Fanar and Oswin by myself."

"Oh, no, you won't!" Rhona pushed an arm out from under the cover and waved at the door. "Close the door."

"I knew that would get you up." Contessa closed the door. "Computer, lights at seventy." The ceiling glowed brighter. "I'm going to change. You have to try the water. It's really good." She slid a wall panel at the foot of her bed to the side, revealing a built-in closet. The top was just above her head and the bottom at her waist. It measured a meter wide and a little over a half meter deep. Below the closet, four drawers occupied the rest of the space to the floor. One half of the closet and two of the drawers contained her items.

"I want the red pants and cream top," Rhona said.

Contessa grabbed the outfit and handed it back to her sister, taking the dark green pants and shirt for herself. After changing, she folded her nightgown and put it at the foot of her bed. She glanced at her sister's bed and shook her head. *Will she ever grow up?* The cover was a twisted lump with the pajamas wadded into a ball next to the pillow.

The boys were waiting just inside the maintenance building, talking to Kellach, when Contessa and Rhona arrived.

"Credits?" Fanar looked confused. "I didn't know we had to pay. C'mon, Kellach. This is a new world. We should let go of such a system."

Kellach snorted and leaned forward. "Your dad insisted. Me? I thought the same as you." He glanced at Contessa and Rhona and stretched his neck to see if anyone else was around. After punching the display with his meaty fingers, he took note of something and motioned with his hand. "Follow me."

Contessa and Rhona followed Fanar and Oswin into the bowels of the large building. An office with a windowed door lay immediately behind the front room. Past that were open rooms used to store equipment. Shelves were loaded with solar panels and other items Contessa couldn't identify. The hallway ended halfway through the building at an open space where the rovers were kept along with other items too big for the storage rooms.

"You're lucky today. We have one left." Kellach led the group to the lone rover by the bay door. It was an oddly designed thing with two articulated segments. The front passenger segment had four tires. The back segment, with only two tires, had a place for cargo and more passengers. Kellach patted the rover. "Where you going?"

Oswin grinned. "Past the lake. Thought we would take in the waterfall and maybe explore the forest."

Kellach nodded. "That's a long trip. You sure you're up to it?"

"Of course we are," Fanar said. "We're still young enough to handle the long days."

"Me? I don't think I could. Besides the waterfall and trees, there isn't much down there." Kellach walked around the rover, opened the opposite door, and tapped in an access code so they could use the rover. "Hmm, looks like the maps aren't loaded on this one. Drive it over to the IQ building to load them in. You know how to do that, right?"

One of the soldiers emerged from the hallway. He seemed to be about twenty-one or twenty-two. Contessa's age. Muscular, but not too much. His brown eyes met hers briefly before he looked at Kellach. "If they don't, I do."

A flash of irritation crossed Kellach's face before he smiled and swiveled around to face the soldier. "Randy."

"I've been assigned to go with them."

Fanar frowned at Randy and answered Kellach. "I know how, but I thought they were going to load in the waterfall so the vehicle could drive itself there."

Kellach shrugged. "It's probably in the map data load. If not, you got the training."

Rhona raced to the vehicle, opened the door to climb in, and stopped. "This has a steering wheel."

Fanar laughed. "Yeah. We had to take driving lessons as part of our training."

Randy stiffly approached the vehicle, his head held high. "I'll do the driving."

"What? No!" Fanar exclaimed. "This is the first chance I've had to drive one of these."

Contessa couldn't believe her ears. *Is he out of his mind? Why would he argue with a soldier?*

"I'm here for your safety, so I'm driving." Randy climbed into the seat. Fanar's face grew dark, and he was opening his mouth to say something when Oswin stepped up. "Randy, right? I think what Fanar is saying is he spent a lot of time learning how to drive this thing back on Earth and he really doesn't want that time to go to waste. After all, you've gotten to drive this already, haven't you?"

The soldier looked between Fanar and Oswin. "Fine, but I will take overif I feel the need." He slid over to the opposite side.

Oswin climbed in and moved to the middle so Fanar could get behind the steering wheel. The wide vehicle could comfortably seat three people across. Beaming, Rhona clambered in behind Fanar.

Contessa climbed in, forcing Rhona to the other side before closing the door. While situating herself, the vehicle lurched forward, causing her to fall into Rhona's lap.

Rhona whooped. Laughing, she pushed Contessa up. "Careful, Sis! We're in for a ride!"

"The IQ building?" Oswin asked. "I was thinking of something like 'The Brain.'"

Fanar spun the steering wheel as they exited the east side of the building. Contessa gripped the handle above the door. Thankfully, they were the only ones on the road as they careened onto it. *This is a ride, all right.*

*Maybe I shouldn't have come.*

"Take it easy!" Randy yelled out.

"The Brain?" Fanar straightened out the path of the rover. He rolled his head around. "Not bad, not bad. Maybe 'Einstein' or 'Mastermind'? Oh, wait! I have it. 'The Oracle.'"

"The Oracle?" Oswin seemed to weigh the words then smiled. "I like it. A god-like source of knowledge and forecasting. That could work."

Rhona laughed. "You guys are hilarious."

They slowed as they approached the building. Several people were near the building synchronizing their uplinks to the core. They would do their journal entry on the uplink and sync when they had an opportunity. Because the uplinks were designed to always be connected to the linknet and its servers, local storage was limited to conserve resources. Many people gave up with the uplinks and simply went inside the building to use terminals.

Fanar stopped the vehicle. "Let's see if this is close enough." He tapped the console to the right of the steering wheel.

"You need to get closer. Still no signal," Oswin said.

Fanar attempted to move toward the back of the building but had to stop. Diggers were making the site ready for the portal generator that was scheduled to be delivered in a few days. An addition to the computer building would be constructed to house the equipment used in opening the portal.

"Here we go; there's a space we can move into," Fanar said and inched the rover up to the building. After stopping, he tapped the screen and nodded before touching more buttons. "Here it is. It would be nice if they could figure out why our radio signals don't work here."

Contessa used the time to study Randy some more. He seemed unsure of himself, almost like he had something to prove. *Maybe because he's the youngest of the soldiers.* He watched Fanar and Oswin as they worked the console. *Such lovely brown eyes. Shame he isn't just a little older.*

Rhona jumped forward and leaned against the front seat, interrupting Contessa's thoughts. "Let's see the waterfall."

Oswin pointed to the bottom of the lake on the display. "It's here." He tapped some more buttons, and an image appeared.

"Nice. Let's go!"

Fanar was reaching over to touch the center console when he heard a tap on the window. His dad stood outside.

His dad waited for Fanar to lower the window. "Off to explore?"

"Yeah," Fanar replied. "We're going to see the waterfall."

"That should be nice," his dad said. He noticed Randy and nodded at him.

Fanar's father leaned in next to Fanar's ear and whispered, "I'm depending on you to keep everyone safe. Remember, you do the right thing, and everything will turn out right." He stepped back and in a louder voice addressed everyone inside the rover. "You kids have fun and be careful."

"We will," Oswin and the girls replied. Fanar watched his father depart before focusing his attention again to the center console. "Let me put it into automatic mode." He tapped another button, and the vehicle moved itself back from the building, turned around, then propelled forward around the settlement toward the lake.

"Doesn't satnav use radio? If we don't have radio, how does this work?" Contessa asked.

Oswin shrugged. "Fanar probably knows. He's into all that technology stuff."

"The rovers don't use satnav," Fanar instructed. "Instead, they keep track of their position by the compass reading and tire rotations. It's surprisingly accurate."

It took about ten minutes to get to the lake. Oswin shifted in his seat.

"Fanar and I were here last week. You get a chance to visit the lake yet?"

Contessa shook her head. "No. I was hoping to the other day, but then something came up. You walked?"

He grinned. "Yep. Only took a little more than an hour."

The rover angled away from the lake to move around the eastern edge. Rhona stared out the window at the water going past. "It's so beautiful."

Contessa didn't think she'd ever seen anything so beautiful. The sunlight reflected off the ripples of the lake like a thousand tiny stars. She lost time in those pinpoints of brilliance and eventually noticed the lake's surface rising as they descended the mountain. Soon the lake disappeared from sight. She sighed and stared ahead to the forest in the distance with a dark ribbon meandering its way through.

On their left, opposite of the rocky mountainside, purple shrubs the size of small dogs peppered the landscape. "They look like leafy mushrooms, don't they?" She looked over her shoulder to Rhona for confirmation.

Rhona briefly met her eyes and nodded.

The ride continued in silence, giving the impression talking would somehow disturb the scene. The electric motors of the rover hummed, and the tires crunched on loose stones as the vehicle sped along. Contessa must have fallen asleep because she awoke to a low rumble that progressively got louder as they proceeded toward the forest. When they passed the edge of the mountain cliff, reverberations of thousands of kiloliters of water per second hitting the rocks submersed in the riverhead assaulted her senses.

"Did you fall asleep?" Rhona asked. She laughed as the rover came to a stop. "I can't believe you fell asleep!"

Contessa ignored her and got out, quickly covering her ears. A fine mist, like a fog, continually rolled out from the base of the waterfall. The water collected on her arms, forming thick drops that ran down to drip off her elbows, chilling her arms.

Oswin motioned for everyone to get back into the rover.

"Wow! I never knew a waterfall could be so loud," Contessa said after they got back inside.

With a lurch of the rover, Fanar manually drove a few hundred meters away from the waterfall and stopped again. Oswin got out and approached the river. The waterfall could still be clearly seen and heard, but it wasn't nearly as loud. Contessa left the rover and walked with Rhona and Fanar. Randy stood near the rover and narrowed his eyes as he studied the area.

"Oh, look!" Fanar pointed to a furry grey creature the size of a mouse with six legs and no tail sitting near the water. With a squeak, it skittered into a nearby bush that grew around a narrow tree. He approached and patted the tree. "This would make a good target."

Contessa scowled. "Target?"

Oswin answered, "Oh, yeah. Fanar's an expert marksman."

"I thought guns were outlawed," Rhona said.

Contessa glanced at Randy, who stood impassively, his rifle slung over his shoulder.

Fanar grinned. "Not guns. Throwing knives and throwing stars. I've been doing it a long time. I packed my set; I'll have to find some time to use them."

"Well, I don't know about throwing knives, but I know this water is calling me!" Rhona took off her shoes and waded into the water. Soon Oswin and Fanar were beside her. She squealed and held her hands up to shield her face when Fanar splashed her.

Fanar laughed and splashed her a couple more times before he turned and smiled at Contessa. "C'mon in. It's not that cold."

Contessa drew near to the water's edge. It appeared cold. "No, I don't swim."

"It's hardly swimming." He dipped his hand into the water and hurled some toward her.

She shied away. She didn't feel like getting wet. "Stop it! You're getting me wet!"

He scoffed. "You're already wet from the waterfall. Seriously, you'll be fine." He sent a series of splashes toward her.

"I said no!" Furious, she spun around and stormed away.

"Contessa—" Fanar called out to her.

"Just let her go. She'll be okay in a few." Rhona's scornful voice chased after her. She'd never understood Contessa's dislike of the water.

Contessa crossed her arms and strode along the river past some bushes.

A part of her heard Randy yelling out, "Don't go far."

Why did Fanar make her so mad? He was right, of course; it wasn't swimming. She examined her clothes. The areas he'd splashed were almost indistinguishable from where water had soaked in by the waterfall. The water mottled the green fabric with dark wet patches. *I could have joined them, but I just didn't want to. What's wrong with that?*

She looked back. She couldn't see them anymore. Several trees and bushes were in the way. She must have gone farther than she'd realized.

The river carved itself into the ground, flowing a half meter below the edge of the land.

Oswin never had said whether he'd seen any fish. She and Rhona hadn't gotten any of the training the boys had received nor had they seen any of the photos from the drones, so she felt at a disadvantage. She stopped close to the edge and peered into the water.

*Is that why I'm upset? Of course, it could be because of Fanar's accusations against Daddy.* She immediately dismissed the notion. No. In a twisted way, his reasoning made some sense. After his apology, she couldn't help but forgive him. She had been off-balance after her dad had accepted the job to come to Niton. Her daily routine and structure was all but forgotten. *I need to find my routine here.*

Some movement upstream caught her attention. She jerked her head to look, shifting her weight onto her left foot. The few centimeters of dirt by her foot collapsed into the water. She pulled back, but her right foot slipped. She yelled out in a panic as she went feet first into the water, hitting her head on the ground as gravity pulled her down.

# CHAPTER SEVEN

## Sickness

**C**ontessa's eyes snapped open to a world of water. Instinctively, she pushed down with her hands, and after her head popped out of the water, she gasped for air. She reached down with her toes but couldn't feel the bottom. Her heart pounding, she searched for the shore. Fear wrapped its icy grip around her as the water twirled her around and moved her with the current, pulling her quickly down. She pushed her hands down again.

"Help!" Her chin quivered as she blurted out her cry and gasped again for air. Her clothes clung to her body, and her muscles cramped. The world spun around her. She slipped under the surface. She kicked. Hard. Her face met the warm air, and she sucked in a breath. Still inhaling as she went down, she took in water and choked. Now in full panic, she pushed down with her arms and broke the surface, coughing and straining for air. A breath. "Help!" She could barely get out the scream before more water entered her mouth. The water twirled her around, making the land appear and disappear. She got pulled under again before she could catch her breath. She pushed with her legs but remained under water. Her lungs burned for oxygen; she couldn't hold her breath any longer. *Oh, God, I'm going to die.* Her salty tears mixed with the river as she exhaled.

She slammed into a boulder and turned. With darkness closing in, she opened her mouth and took in—a breath of air. A feeling of ground against her feet. She found thought difficult as she struggled to maintain consciousness. *Is that the bottom? I must be near the side.* She could feel more of the sandy bottom on her feet. If she could just get onto the land.

A dark shadow loomed over her, a hideous face and bad breath, and suddenly ropes were around her body, pulling her off her feet. *No!* Darkness overtook her.

"Tess!" Contessa heard her sister's pained voice as hands rocked her body. "Please. Tess!" Contessa opened her eyes to see Rhona's tear-streaked face over hers. Next to her stood Randy. Rhona barked out a laugh. She gripped Contessa in a hug, her curly hair getting into Contessa's face.

Contessa coughed and sat up. Fanar and Oswin stood next to her, their faces reflecting the terror she'd experienced earlier. The ground was brown and wet, but it wasn't dirt. Some sort of very short spongy covering. Sunlight burst through the plum-colored forest canopy in patches, revealing the rover between two trees with water dripping from its underside.

Contessa coughed. "Where am I?"

Fanar offered his hand to help her stand, noticed an insect had dropped on it from a tree overhead, and brushed it off before offering his hand again. "About two kilometers down the river. Contessa, I am so sorry."

She looked at his hand and accepted it. "Where did he go?" She wobbled onto her feet and nearly fell against a brown boulder next to her. The sound of the river and the distant roar of the waterfall were the only sounds.

"Who?" Fanar searched the area. "The man trying to take me."

Randy gripped his rifle and peered into the trees. "Someone was taking you?"

"Yeah, and he had really bad breath."

"Are you okay? I didn't know you can't swim," Fanar said.

"It's my fault. I shouldn't have gone on my own." She studied their anxious faces. "You didn't see him? I thought—I almost drowned. Then someone put ropes around me."

Rhona put her hand on Contessa's shoulder. "Nobody was here." Randy shook his head. "We didn't see anyone."

Had she imagined it? She was sure someone had attempted to abduct her. But they hadn't seen anyone. Maybe he'd pulled her out and left? But why would he do that? Nothing about it made sense. Her gaze went to the dripping vehicle. "Why is the rover so wet?"

Oswin sighed. "We had to come across the river. When you didn't come back, we searched for you but couldn't find you, so we went back for the rover. Your sister's the one that saw you lying on the ground. She was so scared. We all were."

"These two were talking about trees. Good thing I kept my eyes open," Rhona said.

"You must have scared him off crossing the river," Contessa said. Her head had cleared.

"We need to go back," Randy said. After a second, he continued, "We'll report what happened here."

Contessa really wanted to go back. This was more adventure than she ever wanted. She and Rhona joined the boys in the rover. Randy climbed behind the steering wheel. He was now driving. Fanar sat between him and Oswin.

"We'll go back the way we came." Randy turned the vehicle around.

The rover submerged into the water until the surface sat just a few centimeters below the top of the doors.

Contessa focused on the floor, expecting water to be pouring in; instead, it remained dry. "Are we floating?"

"Yeah! This thing is stormin!" Rhona beamed.

After a few minutes, they reached the other side and drove up onto the land, scattering several of the tiny grey creatures before Randy aimed the rover toward the falls.

Rhona watched the creatures and frowned. "Those little mousey things are all we've seen. What are they called?"

"Hexaminutoides. Fanar and I just call them mice," Oswin said. She smiled. "Couldn't come up with a better name?"

He shrugged. "Nope."

Contessa closed her eyes. Her body felt so heavy. Could it be because of the stronger gravity? "You okay, Tess?"

She opened her eyes to find Rhona staring at her. She waved her off. "Yeah, just tired."

Without warning, the rover came to a stop. "Why you stopping?" Oswin asked.

Randy growled. "I'm not. It just died." He tapped the console. It remained dark.

"What's wrong?" Rhona peered over the seat.

Fanar tapped the console again. "It isn't responding."

"Are the batteries dead? You were watching the charge, right?" Oswin asked.

Fanar responded, "Yes I was watching! It was at eighty-five percent when we stopped at the falls." In a huff, he motioned for Oswin to get out and walked to the back segment.

"What're you getting?" Oswin called out. "Solar panels."

Oswin peered at the darkening sky. "I don't think those will do any good."

"Do you have a better idea?" Fanar wrestled the panels out of the storage area. "Maybe we can get enough charge to turn it on and see what's wrong."

Randy helped attach the panels.

Contessa shivered. "Wish I'd brought a coat."

"You're cold?" Rhona frowned. "It must be at least twenty-four degrees."

Oswin appeared next to her with a thermal blanket. "There are only two, so we might have to share if it gets cold."

"Thanks." Contessa took it and wrapped it around herself. She would have preferred something heavier. It seemed so light and useless. How could they share such a tiny thing?

Fanar plopped into the front seat and crossed his arms. "This is so frustrating. The battery pack isn't taking a charge. My guess is there's a fault in the battery pack. Probably a short."

"Did the water damage it? Can you fix it?" Oswin asked. "I don't think so. Not here."

"What now?" Rhona squirmed in her seat. "We gonna be stuck here all night?"

Randy sat behind the steering wheel. "Kellach knows where we are, so they should be able to easily find us. If we remain calm, all should be well." His demeanor had changed. To Contessa, he didn't seem so… haughty.

Contessa closed her eyes. The shivering had stopped, but she was still cold. The sound of the river rushing nearby provided some comfort. She might not be able to swim, but something about the sound of running water soothed her.

"Here they are." The voice seeped through Contessa's consciousness. It sounded familiar. *Kellach?*

"Thank God they're okay," Dr. Monier said. "Fanar?"

Contessa opened her eyes and immediately closed them as a beam of light blinded her. When the light moved on, she opened them again. Her body was pressed against the side door with Rhona lying against her, snoring. The light of the three moons made it easy to see Kellach standing next to her and Dr. Monier peering intently into the rover.

Fanar stirred. "Dad? What are you doing here?" "Finding you. What happened?"

Randy waved his hands at the center console and gave his report. "It died. Fanar attached the solar panels, but there's a problem with the batteries holding a charge."

"Hmmm?" Rhona sat up and covered a yawn. "Oh, good. They found us."

Kellach opened the driver's door. "You mind if I take a look?"

"Yeah, sure." Randy climbed out of the vehicle with his eyes half open and leaned against the door next to Contessa. Fanar and Oswin got out and watched.

Kellach sat in the driver seat and tapped the console screen. Not getting a response, he pinched the sides of the screen and pulled it out of the dashboard along with a ribbon of wires. He grunted and pushed it back into place. He then climbed out and went around to the back section. After some banging and more grunting, he returned to stand next to Fanar. "Guess we'll need to tow it."

Fanar pulled his shoulders back and addressed his dad. "It isn't my fault.

It said we had eighty-five percent power—"

Dr. Monier held up his hand. "I'm just glad you're okay. Sit tight and we'll pull you guys back."

Kellach moved the other rover into position and attached a tow line.

The sun had risen when they arrived at the settlement. Contessa saw several people at the lake's edge and in the water, watching them.

Fanar turned in his seat and asked her, "How you feeling?"

At least he seemed to be in a better mood. His usual broodiness was wearisome. He had irritated her yesterday with the splashing, but now he seemed more lighthearted. Contessa met his gaze. "Worn out."

"Really? You slept half the way here."

An idea struck her. "Maybe it's the water. I think I drank a couple liters while going down the river."

His face darkened. "I'm sorry."

"It's not like you pushed me in. It was my own clumsiness."

He nodded and lightened up. "You were the first person to get an in-depth look at the river."

*Good, a joke.* She smiled at Rhona. "Jealous?"

Rhona smiled back. "Not this time, Sis. I'll discover something more interesting."

They slowed as they approached the maintenance building and came to a stop inside. Randy mumbled something about reporting back and left.

Contessa clambered out of the rover and stopped. Darkness closed in around her. She reached out and grabbed the top of the door to keep from falling. *What's wrong with me?*

"Whoa, you okay? Here, lean over." Kellach grabbed her arm and gently pushed her forward with his other hand on her back.

Murmurs surrounded her, and the darkness grew until she could no longer see. But she was still conscious and standing. Okay, more leaning on Kellach than standing.

"Do you think it was something in the water?" Oswin asked. "She fell into the river and nearly drowned. She said someone was there, but we didn't see anyone."

"It's probable she ingested some bacteria, but I wouldn't expect this kind of reaction," Dr. Monier said. "Let's get her home. Fanar, run and get Dr. Huff over to her place."

The floor came into focus and then the surrounding area. Contessa took in a deep breath through her nose and slowly let it out through pursed lips. Warmth crept into her legs, and she stood. She didn't need help; she had this.

"Take it easy. Don't move too quickly." Kellach continued to hold her steady.

"I'm okay." She took a step away from him.

Rhona scowled. "Think you can get back home, or should we bring the doctor here?"

The concern on Oswin's face mirrored Rhona's. Dr. Monier stood nearby with an expression difficult to read. *Where's Fanar? Oh, yes, getting the doctor.*

Contessa nodded. "I'm okay." To prove her point, she took several steps to the open bay door and looked back. "See?" She didn't need help. She was the one who provided help.

Dr. Monier nodded and addressed Oswin. "Make sure she gets home safely." He directed his attention to Kellach. "Find out what's wrong with this rover and let me know." He walked over to Contessa, patted her shoulder, and left.

Oswin and Rhona took her home, where Fanar and the doctor were waiting with her mom.

"Are you okay? You're so pale!" Contessa's mom rushed over. "Here, let's get you inside so the doctor can find out what's wrong."

Contessa crossed her arms. "I'm okay, Mom. Just tired." How ironic, telling her mom about being tired. She went inside and dropped into a chair.

Dr. Huff took Contessa back to her bedroom and examined her. "I understand you almost drowned?"

Contessa nodded.

"Did they have to revive you?"

"No. They found me unconscious, but I woke up on my own."

Dr. Huff put the stethoscope against Contessa's back and asked her to breathe. "You sound good. You probably ingested water but didn't breathe any in. You're alert, and you walked here on your own. Some good signs. How's your stomach feel?"

"Awful, like I'm ready to get sick any minute."

"I'm going to draw some blood and then give you a general antibiotic." She put a cuff around Contessa's upper arm and inserted the needle into her arm. After filling the tube, she pulled the needle and wiped the spot with some cream that instantly congealed to form a protective barrier. Dr. Huff then handed Contessa a bottle with ten tablets inside and said she would let them know the findings. She gave further instruction to stay hydrated and get some rest.

Contessa followed Dr. Huff to the sitting room and sat while the doctor spoke to Contessa's mom and left.

"Here, let me get you some water." Her mom stepped toward the kitchen. "No." Contessa didn't want any Nitonian water. "Don't we have any water from Earth?"

"Fanar, I think we have some left, don't we?" Oswin asked Fanar. "I can go—" Fanar started.

"I'll go. I'm the fastest runner here." Oswin smiled and sprinted out the door.

"What's wrong with the water here, honey? You know it's been filtered.

It won't hurt you." Her mom tried to console her.

Contessa shook her head. "I just don't want it now." She'd had plenty of the stuff from the river. If that water had made her feel this way, she didn't want anything to do with it.

After several minutes, Oswin returned with two bottles of water. "It's all we have left, but there might be more elsewhere."

"Thanks," Contessa said as she took one of the bottles. She put one of the tablets provided by the doctor into her mouth and drank half the water. It had a familiar chemical taste, but she hadn't noticed before how strong it was. She hoped there were more bottles from Earth. Maybe in a couple of days, she could trust Nitonian water again.

# CHAPTER EIGHT

## Supplies

*This is Fanar. I never found my mom, so I'm not sure why I'm here. Earth will be opening the portal in two days, and we'll be sending back the water we've been collecting and getting a much smaller version of the portal generator—Oswin and I call it the Return Machine—and some supplies. I asked why the Return Machine didn't come with us and was told we needed the transport space for housing and other necessary supplies to properly survive the first month.*

*Besides exploring, there isn't much for me to do until then. Even exploring was a disaster with Contessa nearly drowning and the rover dying. When we first arrived, Kellach said the planet is too beautiful to touch. Maybe he's right. (System mark: Day 28)*

"**S**o, what are you doing today?" Oswin's voice floated through the dark of the bedroom.

Fanar was lying on his back and staring into the inky nothingness. He opened and closed his eyes, not able to tell the difference.

He waved his hand in front of his face, unable to see it. "I don't know. I can help Kellach with some repairs, but I think he's waiting for supplies to be delivered before he can do anything else. You?"

"My parents want me to help them today. They've collected samples from the lake and want my help with the lab work."

"Yeah? You love that stuff. You going after breakfast?"

"Want to help?"

Fanar chuckled. "Don't think I would be much help. Have you heard anything new about Contessa?"

"No. Want to go over there?"

"Yeah. Hopefully she's feeling better. I don't want to leave with that on my conscience."

"You make it sound like you're going to Earth and not coming back."

Fanar sighed. Not finding his mother weighed on him. She had worked so hard on this project and had been so excited to come with him and his dad. He suspected she'd wanted this trip to be a means of reconciling the differences between his father and himself. He'd never told anyone he had secretly hoped to find her on this trip. "I don't want to talk about it."

The ceiling illumination came on at the lowest level. *Oswin must have sat up in his bed.* "Why are you leaving and why are you not coming back?"

"I just don't feel a need to stay."

"Computer, lights at fifty." The room brightened, and Oswin's face appeared over him. "This is a two-year commitment! You have to stay." "Why? My dad barely talks to me, and what I hoped for didn't happen." Oswin frowned at him. "What did you hope would happen?"

Fanar sighed again. "I don't want to talk about it."

"I'm your best friend for almost as long as I can remember. If you can't talk to me, who can you talk to? You can't leave; it's that simple. I I'm getting a shower."

Fanar knocked on the door to the King residence. It was an odd experience. On Earth, he normally used the uplink to announce his arrival. With the radio frequencies inhibited on this planet, such a practice wasn't possible.

Rhona answered the door. "Hey. C'mon in." She led him into the sitting room.

He and Oswin followed her inside. The place was quiet, with Rhona the only person in sight. "Where's Contessa?" Fanar asked.

"She's in bed. Still not feeling well."

Oswin sat in an empty chair. "Any word from Dr. Huff?" "She's stopping by in a bit to check on her."

Mrs. King entered the room. She appeared tired. "I thought I heard some talking. Morning, boys." Had she been up all night worrying about Contessa? She addressed Rhona. "Tess still in bed?"

"Yeah. I had to put an extra blanket on her." A movement outside caught Rhona's attention. "Doctor's here." She got up to open the door.

Dr. Huff entered and nodded to the boys before turning her attention to Mrs. King. "You look tired, LaKeisha. You doing okay?"

"No worse than normal." She shrugged. "I'll take a nap after you leave."

"You need to be careful with your condition."

*Condition?* She often seemed tired, but Fanar thought that was her normal appearance or that she was possibly overworked.

Mrs. King changed the subject. "What'd you find out about Tess?"

Dr. Huff pulled her lips into a thin line. "She's a little anemic. Also, I can find no evidence of a bacterial infection. I believe a toxin has entered her system."

"How? Was it in the water?"

"I don't know. I don't think it's the water, but we've only been using what's filtered from the lake. Perhaps something downstream—" "What do we do?"

"If we could isolate the toxin, it would help in giving a proper treatment.It would be best if she went back to Earth tomorrow. They have better facilities, and the water samples from the drones may give a clue to the toxin. We have information here in the IQ system, but not all the samples had been analyzed when we left Earth." She paused. "She can get the care she needs. Until I know what we're dealing with, it's best for her."

Mrs. King's eyes widened. "Is it that serious?"

Dr. Huff took in a breath. "I don't think so. She's strong and stable. It's the unknown that worries me more than anything else."

"I'll go with her, Mom." Rhona looked like she was going to cry.

"The three of us will go," her mom answered.

"And leave Dad here by himself? No, you need to be with him. I'll go, and then we'll come back next month when they open the portal from here." Fanar shifted in his seat. *Is this family for real? What family gets along like this?* He clenched his jaw and glared at them in envy. It wasn't fair, them being close while he and his dad could barely speak to one another. He nearly stood and left. Still, he couldn't help feeling guilty about the situation. Rhona had been so excited about coming. He cleared his throat. "If you like, I can take her to Earth when they open the portal. I could stay behind on Earth and check in on her."

Rhona stared at him, her eyes wide and a small smile on her face. Fanar took it to be appreciation. "That is so nice. But like my mom said, we'll go together."

He bobbed his head.

Dr. Huff shifted her bag from one hand to the other. "You mind if I go and check on her?" Without waiting for an answer, she left the room.

Fanar and Oswin stayed a couple more minutes before leaving.

"Did you mean that?" Oswin pushed his limp blond hair away from his eyes as they headed toward the lab where Oswin's parents were working.

"Mean what?"

"Taking Tess back to Earth with you."

"Yeah, well, you know. It's the least I can do after what I did."

"It's not your fault. All you did was splash her. How were you to know she would fall in the river and almost drown?"

"That's pretty much what she said." They stopped outside the lab. "I know you're right."

"Of course I am. I'll see you later." Oswin disappeared inside.

Despite what Oswin said, Fanar couldn't shake the guilt for Contessa's situation. His capacity for any cheerfulness at zero, he trudged up the road past the housing units until he got to the main road and walked around toward the maintenance building. He was about to enter when he saw Mr. King leaving in a rover with some other men.

"And so it begins," said Kellach from behind him in his usual gravelly voice.

Fanar jumped and turned to see Kellach standing in the doorway, watching the rover depart. "Oh, hi, Kellach." He glanced at the rover. "They're going to start the mining?"

Kellach seemed sad. "Yeah. First they'll release rock and dust into the air to remove the rare mineral in chunks along with the unwanted rock. Then they'll wash it down with water and caustic chemicals to clean it off, with the runoff spreading across the landscape and into the lake to be taken downriver. Then they'll store the mineral for shipment back to Earth for further purification." His eyes moved to Fanar and held his gaze as if measuring his worth. "How do you feel about it?"

Fanar shifted his weight. Kellach's mood often matched his own. Perhaps that was why Fanar was comfortable around him. Today Kellach seemed more critical. Fanar carefully selected his words. "I'm sure it won't be that bad. My dad's a careful person."

Kellach grunted and went inside. "What can I do for you?"

Fanar followed. "I came to see if you need help with anything today."

"You did a good job cobbling the houses together. Even calibrated some items before I had to tell you to." He opened the door behind the desk and led Fanar back. "Let's see what you make of this."

"I didn't even know there would be mining until a couple days before coming to Niton. I thought we came here for water and growing plants."

Kellach opened a door and entered a small room. Inside was a metal workbench with a piece of research equipment half assembled.

"A 3-D microscope." Fanar studied the pieces on the benchtop. "What happened?"

"This is one of the items found damaged when we first arrived. It's recently become a priority. The stage is immovable, and one of the lenses is useless."

Fanar turned to Kellach and saw him staring back. "What do you want me to do?" *Surely he doesn't expect me to fix it.*

"Me? I want to see what you can do to fix it."

*How am I supposed to do that?* Sure, he had tinkered with electronics many times in the past, but this was a sophisticated piece of equipment. He took a deep breath. "Okay. Let's see here." Tools lay scattered across the workbench surface. The lens assembly and image processing unit had already been removed and were lying next to the microscope. He picked up a small screwdriver to unscrew the access panel to the inside of the arm. Wrong size. He tried another, and it fit. He pulled out the screws on the neck and underneath and placed the plate to the side. Inside were rechargeable batteries, gears, a motor, and a wire ribbon to the touch panel built into the top of the base. Funny how technology could advance, but for some functions, it came down to the basics. Fanar searched the workbench.

"Whatcha looking for?"

"Voltmeter?"

Kellach opened a drawer above the bench and pulled out a voltmeter.

After testing the battery output and various endpoints, Fanar put the voltmeter down and tapped one of the control buttons on the touch panel. Nothing moved. He accessed additional controls for granularity of movement and tried again. Frowning, he tapped several times on the screen. *Where is it?* He found the test mode button. He tapped it, grabbed the voltmeter, and placed the leads onto the motor's connections. Nothing. The motor was damaged. He looked up.

Kellach scratched his chin and grunted. "Test mode, huh? How'd you know about that?"

"My mom. That's how Oswin and I met, you know. My mom worked with his parents."

"No, I didn't know that. Okay, so what now?"

"It's a specialized stepper motor with a lot of fine wires. Not something you can repair. It has to be replaced."

"So we just toss this old one? Bury it in the ground?"

"No, we recycle—oh." Fanar nodded. With no facility on Niton for recycling, items would need to be stored and sent back to Earth. Similar to the point Kellach had made earlier with the mining. "So it begins."

"So it begins." Kellach reached over and detached the motor. It was mounted in a socket for easy replacement. "They're opening the portal tomorrow from Earth. How do you feel about going? You can get this replaced and take back some data."

"Yeah, I already offered to take Contessa back." "Good."

Kellach strode to the front area with Fanar in tow. He placed his thumb on the face of a drawer. It lit up around his thumb with a green circle and opened. He pulled out a memory disc, about two millimeters thick and a centimeter in diameter. "It has a log of the issues we've been having and needed supplies, among other things." He put it into a case designed to hold a couple dozen discs.

Fanar accepted the case. "Do you want to put the log entries from the Oracle onto it?"

"I've been hearing some people refer to the IQ as the Oracle. I assume you have something to do with that?"

Fanar grinned.

"Cute. Yeah, those are on there too. Give it to Ian Jones. He'll know what to do with it."

The front door opened, and Oswin came in with Rhona.

She put her hands on her hips, her fingers facing back. "We need to deal with my sister. She's refusing to go back to Earth."

Fanar frowned. "What? Why?"

"I don't know. I thought maybe Oswin could talk some sense into her, and he insisted we get you."

# CHAPTER NINE

## Earth

**C**ontessa was sitting on the couch. She seemed a lot better than when Fanar had last seen her. Her face had more color, and she appeared more alert. She snorted when they entered. "Ganging up on me?"

Rhona sat next to her sister.

Fanar sat across from Contessa. He rubbed his forehead before saying,

"We're not ganging up on you." She raised an eyebrow. "Oh?" "How are you feeling?"

"Fine."

Rhona scoffed. "No, you're not." Her voice got louder. "You complained while getting out of bed. Dr. Huff said you should rest and that you should go back to Earth."

"I can't, and you know it." She gave her sister a hard look. "I'm staying."

"You know they'll open the portal again in a month," Fanar said. "It will give you time to fully recover, and then you can come back." "I'm already feeling better. I'll be fine."

"What's all the fuss in here?" Mrs. King entered the room. "I can't even take a nap."

Rhona stood. "Sorry, Mom. We didn't mean to wake you."

Mrs. King sat in the vacated spot next to Contessa. "I was already awake. Contessa, you should be in bed." "I'm fine, Mama."

"Mom, tell her she needs to go back to Earth."

"Why don't you want to go back to Earth?" Mrs. King put her hand on Contessa's leg.

Contessa frowned. "I need to stay here." She lowered her voice. "You need me."

Mrs. King hugged her. "I love the way you watch out for me, but I can take care of myself. I've been doing it since before you were born. You let this fine young man take you and your sister back to Earth, and you get better. You'll be back before you know it."

Contessa remained silent for several seconds. "I don't want to leave. I'm feeling better than yesterday. Even the marks where the rope was put around me have gone away. I don't think I need to go. Maybe they can take a blood sample and come back with a report. That's all they would do if I went, right?"

Her mom held her gaze. "Tell you what. We'll see what Dr. Huff says about your idea."

"Fine. Oswin, will you get Dr. Huff?"

Oswin nodded and pushed himself off the seat.

When Oswin arrived with Dr. Huff, Contessa and her mother had both fallen asleep, leaving Fanar and Rhona talking between themselves.

Rhona woke them. "Dr. Huff is here."

Contessa asked the doctor, "Did Oswin tell you my idea?"

The doctor nodded. "I would rather you go back until we know what this is. However, you are clearly getting better and are determined to stay. I can prepare a data disc along with a blood sample to be taken back. Who knows? Perhaps they can match the toxin while the portal is still open. Fanar, you still willing to go and have this analyzed while the portal remains open? Since they have all the sample data there, they could probably identify it a lot faster than we could." She directed her attention to Oswin. "Why don't you go with him? The Healing Pursuits Lab is just down the street. They're working closely with the institute. You can both go and deliver the package and wait while the computer scans for a match. Maybe we'll get lucky."

"What if one can't be found?" Mrs. King asked.

Dr. Huff smiled. "Right now, we want to limit the impact of whatever she ingested. But she'll probably be okay as long as she doesn't take another swim in the river. It's also useful to know what affected her and have an antidote for other cases that might occur. If they can't match the toxin while they wait, we'll get an answer next time the portal opens." She looked between Fanar and Oswin. "How's that sound?"

Oswin nodded. "Sounds like fun."

Fanar smiled. "Kellach already has a task for me. Soon I'll be a messenger for half the colony." Inside, he grimaced. He was supposed to have come to Niton with his mom and dad. Since she clearly wasn't here and his dad might as well not be, taking Contessa back to Earth would have been an easy way to manipulate his permanent departure from Niton. Now Oswin would be on him about returning.

"The first priority item will be the calibrated wormhole generator." Fanar's dad addressed the dozen people who would be traveling through the portal. Fanar shivered against the cold morning air. He, Oswin, and the others waited on the plain north of the Oracle building where they had first arrived.

"Too cold for ya?" Oswin grinned at him and shifted his jacket.

Fanar rolled his eyes. "You're going to be hot on Earth." "I'll be okay."

"We'll be transporting the water to Earth, and once that's through, we'll receive the wormhole generator," his dad continued. "Then we'll have the requested supplies come back to Niton. Any questions?"

Dr. Yamamoto raised her hand. "Jim, I don't recall if we're expecting the seeds today, or if that's next time."

"We had a set list of items that included the seeds. There will be two platforms ready. However, because of our unexpected issues, I'm personally going through to see what damaged equipment can be replaced immediately." He directed his attention to Fanar. "Kellach says you have a data disc?" After Fanar nodded, he resumed, "Fanar will deliver the list of needed supplies before running an additional errand, and I will coordinate with the team Earth-side. I don't know if the seeds will make it because of this change in plans. Sorry, Aiko."

She waved it off. "I understand. Hopefully they will come, but if not, I still have plenty to keep me busy." "Definitely a lot to keep us busy."

Randy came up between Fanar and Oswin and smiled at Fanar. "This is fun."

Fanar scowled at him. "What's fun? Why are you coming?"

"Going through the portal. It's fun. You start here, and in the blink of an eye, you're someplace else. It's stormin! I was asked to go with you and Oswin to some lab."

Fanar's chest tightened. Going through the portal was definitely not fun. When the portal opened, a single shockwave that could be felt to the bones enveloped him, similar to standing next to a giant kettledrum that someone pounded on with a soft mallet, but without the sound. Odd how it didn't feel like that when they were on Earth. Maybe something about the open area compared to the underground placement on Earth. He stared at the shimmering greyness and wondered if he should try to make this a one-way trip. What was here on Niton to keep him? Clearly his mother wasn't here, his thought of discovering her somewhere unfulfilled. His father didn't have time for him.

After several seconds, one of the four platforms carrying water slowly made its way through to Earth. Dr. Monier moved to the side and led the way past it to Earth.

"Now for the pain," Fanar muttered to himself. Oswin looked at him. "What are you talking about?" "Going through the portal. It hurts."

"I don't feel a thing."

Fanar followed the thin line of others going to Earth past the lumbering platform and stepped through the portal. Like before, he entered a world of white mist and was stricken with a thousand needles.

When the departure room appeared, Fanar moved to the side, wincing in pain. Many people were waiting and descended upon the team.

Randy looked at him, a wide grin on his face, which quickly disappeared.

"You okay?"

Fanar could barely hear him. The agony quickly eased, leaving the familiar tingling caused by proximity to the active portal generator. He nodded at Randy. He watched his dad engage with several people, leaving Fanar to himself. Just more evidence his dad didn't need him. It didn't matter; he didn't need his dad either.

A tall man with a beard stood nearby, acting as if watching for someone.

Fanar asked him, "I'm looking for Ian Jones. You know him?"

The man leaned in and held his hand behind his ear.

Fanar had forgotten about the effect of the portal on sound. He repeated himself with his voice raised.

The man smiled. "That's me."

"Great." He pulled the case out of his pocket. "Kellach has a disc for you." He paused. Inside the case were two discs. One for Ian and the other with information regarding Contessa's condition. "We'll have to see which one it is."

"He would have encrypted the one to me for safe delivery. You mind?" He held his hand out for the case. Fanar studied him as the man checked the discs. He seemed familiar, but Fanar wasn't sure from where. Ian pulled a tray out from the bottom, placed one of the discs into it, and slid the tray back into the case. A blue dot appeared on the top of the case with some words below describing the contents. He pulled the tray out and replaced the disc with the other. A red dot appeared and the description showed Ian's name. "This one's mine. Thanks." He gave back the case, took his disc, and left.

"That's odd." Oswin took off his jacket.

"Yeah, I thought so too. Why would Kellach need to encrypt the disc like that?"

"Well, that too. But I was talking about your dad."

"My dad?"

"Yeah. I thought he would be talking to you by now."

Fanar glanced at the countdown timer on the wall. Three hours and forty- five minutes remaining. Fifteen minutes had already passed. He glanced at his dad and saw men in suits talking to him. The same ones that had threatened to move the program to another country before the colonists departed for Niton. *Why are they here?*

Fanar and Oswin entered the hallway, Randy leading the way. "That's my dad." Fanar's voice echoed back off the wall with more volume than he'd intended. With the door closed behind them, sounds were normal again. He lowered his voice. "He's been like that since Mom—" He paused. He wanted to say his mom had gone missing. This past month, that small bit of hope had died inside of himself. She was truly gone. "Since my mom died," he finished.

"I guess you weren't exaggerating. I guess I never realized how much he avoids you. So going through the portal really hurts? You seem fine."

Fanar rubbed his arms. "Yeah, for a millisecond it feels like tiny needles pricking me all over, and then it's gone."

"Weird," Randy replied. Oswin nodded in agreement.

Fanar stopped and tapped his uplink. "How do I get to the Healing Pursuits medical lab?"

The male voice gave an address, and after Fanar said he would like to be guided, an orange circle appeared on the floor around his feet. It flashed two times before it moved forward a meter and stopped.

Fanar grinned at Oswin and Randy. "It's so nice to be back where technology works."

Oswin held up his bare wrist with a smile. "I forgot mine."

They followed the marker down the corridor and went up to the ground level. As they headed toward the exit, a voice called out behind them. "Fanar! Oswin!" They turned to see Mr. García, the trainer they'd seen almost every day while preparing for the trip. He approached them. "I'm surprised to see you so soon."

Randy stayed in place while Fanar and Oswin met with Mr. García. Oswin smiled. "We're just here to run an errand, then it's back to Niton." "So? What's it like? I bet the lake is even more beautiful in person. Did you see the falls? What about the heavy gravity?"

"The lake and falls are indescribably beautiful," Fanar said. "We took a trip down to the falls and did a little exploring." He paused. Should he mention Contessa? "The heavier gravity takes some getting used to. I got tired easily at first. Coming back, I feel like Superman." He grinned.

"I still can't get used to the longer days," Oswin said. "Awake while it's light some days, awake while it's dark others. It's crazy."

Mr. García chuckled. "I bet. I wish I could join you, but we're getting ready here for the next world. Recruiting people, preparing the sample drones, ordering supplies. After this, you should be self-sufficient. Maybe it isn't so bad for me. After we research and colonize the other eight, I can pick my favorite." He winked.

Fanar laughed. "You guys. Earth isn't that bad, is it?"

"No. It's just exciting to explore new worlds. Like when the pioneers first explored North America," Mr. García said.

"Weren't most of them killed by animals, disease, and the natives?" Oswin shook his head. "Fanar, you just don't get it."

Mr. García pointed at the pulsing circle on the floor. "Where you guys headed?"

Fanar glanced at Oswin, feeling smug. "A perfect example of my point." He looked back at Mr. García. "While exploring the falls, Contessa King fell into the river and nearly drowned. Then she got sick. Dr. Huff thinks it's a toxin but can't identify it."

Mr. García's smile disappeared. "That's awful. Is she okay?"

"She's better. We were supposed to bring her back for observation, but she refused to come." Fanar couldn't keep the agitation out of his voice.

Oswin held up the box he'd brought. "We're going to the lab to have a sample of her blood analyzed. Maybe in the next couple of hours they can match something in it to the samples that were collected."

Mr. García's eyes moved back and forth from Oswin to Fanar. "So you're going there and then going back to Niton? You'd better go, then. I'll catch up with you some other time. It's not like you're completely isolated there." He shook their hands. "I hope you find something to help the young lady." He continued down a different hallway.

Fanar glanced at his uplink. Only three and a half hours remaining. The guidance circle continued onto the sidewalk outside the building. Fifty meters of artificial turf separated the building from the perimeter fence. They followed the guidance circle down the sidewalk, through the gatehouse, and past the fence, where a dozen picketers held vidscreens saying things like 'Don't spoil other worlds' or 'Fix Earth first.' Some were Greeners. Others were sympathetic to their cause. Seeing Fanar, Oswin, and Randy come through the gated area, the crowd rushed up and yelled incoherently. Randy screamed for people to stay back. Fanar winced and pushed his way through. Body odor mixed with the smell of cleansers and deodorants assaulted his senses. How quickly he had gotten used to a daily shower because of the fresh water pumped to the colony. After getting through, he looked back and saw Oswin right behind him, so he continued.

After following the orange circle two more blocks, Fanar stopped. "How far is this place?"

"Dr. Huff said it was just a couple of blocks," Oswin replied.

Fanar tapped his uplink. "Show me a map to our destination." A 3D display projected over the back of his hand indicating his current position and destination. "That's a kilometer away! We'd better hurry." They picked up their pace and continued to follow the circle.

Oswin loped alongside. "Fanar, are you okay?"

"What? Yeah, I guess so. Why?"

"You said you aren't coming back, and today is the first time you've admitted that your mom died." He paused. "Why aren't you coming back?" Did Oswin really have to bring this up? Couldn't he just leave well enough alone? Fanar sighed as tears stung his eyes. When he was told his mom had disappeared, he'd believed it was a mistake. He shifted from excited energy in finding her to depression. His girlfriend at the time left soon after that, saying she didn't know how to deal with his moodiness. His mom had left him, his girlfriend had left him. In many ways, his father had left him. What was on Niton for him? "Fanar?"

"I—I don't know."

"Your mom would have wanted you to do this trip with your dad. You need to come back with me. We're a team, you and me. Who else will help me with my new aquarium?"

"We're all a team," Randy interjected.

Fanar glanced at Randy and focused again on Oswin. "You didn't tell me about a new aquarium. Where would you put it?"

"I don't know. And with all the new equipment arriving, Kellach will need your help. What will you do here? Where would you go?"

Fanar stared at Oswin. Of course Oswin was right. Even if he couldn't find his mom on Niton, he should go back with Oswin and do as he'd promised. Here, he could go live with an uncle, but nothing on Earth waited for him either. At least on Niton he had Oswin. After several seconds, he admitted to himself that his father was there too. He didn't say it, but he had made his decision. He would return to Niton with Oswin. They soon arrived at the lab.

The anteroom was narrow and long with no furniture. Randy walked toward the door at the long end and was reaching for the handle when it opened. He pulled his hand back.

A pear-shaped man in a white lab coat pulled the mask on his face down. Fanar wondered how the man had squeezed through the door. "May I help you gentlemen?"

Fanar dug out the data disc container. "Dr. Huff is hoping you guys can analyze this blood"—he nodded toward Oswin, who held up the case—"and identify a toxin based on the samples taken from Niton."

"You're back from Niton?"

Fanar nodded.

"You'd better get in here." The large man backed into the open space to let them in. "Let me see what you have. How much time before the wormhole closes?"

Fanar glanced at his uplink again. "Two hours and forty-five minutes."

The man shook his head. "Give me what you have and help yourselves to some water." After receiving the items, he pointed to some chairs and muttered about impossible tasks in a dangerous situation as he squeezed through a doorway into the next room.

Oswin pointed at the vidscreen on the wall. "Look, it's your dad."

They listened as Fanar's dad spoke to the reporters, telling them it appeared the Greeners had affected their supplies and that everyone was fine. Without naming Contessa, he mentioned a sick member of the colony quickly recovering from an unknown substance on the planet. He continued, "This endeavor to enrich our planet with fresh resources is more than adequate; it's a great success."

Off to the side, the men in suits were frowning.

Hearing his dad talk, Fanar could almost believe all was going well on Niton.

# CHAPTER TEN

## Rebound

**F**anar pulled a couple of water bottles out of the chiller. He handed one to Oswin. Seeing Randy standing there, he grabbed another and passed it to him. Fanar's uplink buzzed, and a flash of red passed across the screen. *Who's linking me with an emergency?* He tapped the uplink.

"Mr. García?"

"Fanar. You need to hurry back. I tried to link Oswin but couldn't get through."

"He forgot his uplink on Niton. We'll leave in a couple hours so they have time to match that toxin."

"You have maybe an hour before that portal closes. Word got out to the Greeners about people coming back from Niton. Someone got inside, somehow, and damaged the power cells, and now the reserves are low."

"Okay, Mr. García. We're on our way." Fanar slapped the uplink to close the channel and focused on Oswin, his eyes wide.

Oswin returned his look and followed Randy toward the entrance.

Fanar started to follow when an idea hit him, and he ran to the inside doorway. He could see the man who had greeted them standing at a terminal. "Mister?"

The man turned toward Fanar. "Yes?"

"There's an emergency, and we have to go now. But do you have a replacement motor for, um—"

"The Quasar 3D microscope model 4R201," Oswin said and smiled at Fanar.

The man nodded and waddled over to a wall cabinet and opened a drawer. He pulled out the replacement motor. "It's for the 200, but the 201 uses the same motor. Only spare we have." He walked over to Fanar and handed it to him.

"Thanks." He sprinted toward the exit.

"What about this toxin search?" the man yelled after them.

"We'll try to identify it on Niton," Oswin yelled back as he ran through the door behind Fanar.

Fanar quickly fell behind Randy. He had to get back. With his decision made to return, he didn't want to be stuck here. After several blocks, he was breathing hard. He followed Randy around the corner and almost ran into him. "Why'd you—?" He looked past and saw why. When they'd left the complex, there had been about a dozen protesters. Now there where close to fifty. Police officers stood at the gate entrance.

How were they going to get through that? Fanar glanced at his uplink. It had taken fifteen minutes to get here. Angry shouts were directed at the police officers. Most pedestrians passed on the other side of the road. Some stopped to read the messages on the displays. Fanar could feel the energy of the mob from where he stood.

"Maybe if we go around them and run along the fence?" Oswin motioned with his hand toward the fence.

Fanar's uplink chirped. He tapped it to answer.

The face of Mr. García appeared. "Where are you guys?" "We're down the block trying to figure out how to get through."

"I'll tell the officers to expect you. They'll help you get through."

"Okay." Fanar slapped the uplink and moved up next to Oswin and Randy. "You ready?"

Oswin nodded. "Let's go."

Randy led them down the sidewalk and stopped across the from the gated entrance. The building took up several blocks of surface area, but there were only a couple of entrances to maintain security. Fanar waved his hand and made eye contact with one of the officers. The man talked with another officer, and together they worked their way through the crowd and came across the street.

"You Fanar?" When Fanar nodded, the officer looked at Oswin. "And Oswin?" Oswin bobbed his head in agreement. The officer shifted his gaze to Randy.

"Randy. I could have gotten them across."

The officer replied, "It would have been easier if you had gone the long way around, but I understand time is critical. Keep close and we'll get you inside."

Fanar winced as a cacophony of voices assaulted his ears. One officer led the way while the other officer followed, sandwiching them. They inched their way through the crowd with the officers yelling for people to step back. The faces opposing them were twisted with anger. Fanar's heart pounded, his breaths more rapid. Only ten more meters. His muscles tensed. The lead officer forced someone to the right so they could get past. Two other protesters rushed in to fill the space and blocked their progress.

"Move!" Fanar yelled and shoved the protesters aside, amazed at how easily he could. He charged ahead, pushing others aside, Randy doing the same until they reached the clearing by the gate where six other officers stood guard. One opened the gate to let them inside. They dashed the fifty meters down the sidewalk to the building and then to the elevator.

"Wow, you really let them have it," Oswin said.

Fanar briefly hung his head. "Sorry, I panicked and had to force my way through."

"No need to apologize. Just surprised to see such force. And you didn't fall too far behind on the way here."

"I guess it's an effect of living for a month on Niton."

After going down the elevator, they raced through the hallways and entered the departure room. The back end of a platform disappeared into the portal, another poised nearby. A glance at the timer revealed nothing. It had been turned off. They had no idea how long it was going to stay open.

"There you are!" Mr. García shouted from the doorway. "Run! I don't know how long it's going to hold."

Fanar pumped his legs. He couldn't find his mom, but maybe he could find a way to help Contessa. His father stood at the opening of the portal, eyebrows furled as he combed his fingers through his hair. Seeing Fanar, he dropped his hand. His shoulders and face visibly relaxed. He waved Fanar forward and, when Fanar got close, disappeared through the portal.

Oswin sprinted ahead and stopped at the opening. "C'mon!"

Randy bolted past Oswin. A second later, Fanar overtook Oswin, who slapped his back and fell in right next to him into the void. Just as the two of them entered, an explosion shoved him forward, and all went dark.

# CHAPTER ELEVEN

## Sampling

*Log entry for Contessa. Oswin and Fanar didn't come back yesterday. I guess they decided to stay on Earth. I probably shouldn't have come out today because I'm still weak, but Rhona is taking care of me. (System mark: Day 31)*

*s this how Mom always feels? How can she live like this?* Contessa sat on the chair with a blanket on her. The chill had left a couple of days ago, but the blanket comforted her. A glass of water sat on the table next to her, untouched. No water from Earth could be found in Landing. The colonists had started calling it that. Landing. She heard once that it was originally to be called Aryana. Were Fanar and Oswin responsible for the new name?

"It doesn't make sense. They were supposed to come back with the analysis. And why did they close the portal early?" Rhona was still her animated self. She waved her arms about as she sat on the couch opposite Contessa. "You can only imagine the questions Oswin's parents had for

Dr. Monier about it. That soldier has no idea what happened to them, said they were right on his heels. Now they have to wait another month before Earth opens the portal and they can get some answers."

"Can't we open a portal to Earth with the equipment that came through?" "It will take them several days to get it properly assembled and then it will be a month before enough energy is stored. I'm hearing some say Earth is going to open the portal one last time, and after that, we will be the ones controlling travel to and from Earth. Others say this was the last time because they'll be focusing on other worlds."

Contessa nodded. "I wonder how many worlds they're going to colonize."

"I don't know." Rhona's eyes landed on the water. "You're going to get dehydrated. Everyone else is drinking it and nobody is getting sick."

Contessa's dry mouth longed for the water. Logically, she knew it would be safe. But her emotional response to the near drowning and subsequent sickness made her hesitate.

"Don't make me get Dr. Huff. You heard what she said this morning about hooking up an IV if you get dehydrated."

Contessa smiled. Her sister would do it if she didn't drink soon. She brought the glass to her mouth. Over the rim, her sister's eyes met her own. Watching. Daring her to take a drink. Contessa took a sip. "Satisfied?"

"No. You have to drink three of those glasses today." Contessa took a full swallow and put the glass down. Rhona smiled. "See? It isn't so bad."

No, it wasn't.

Rhona continued, "Dr. Yamamoto is going to get water and plant samples from the river the day after tomorrow."

"Past the falls?"

"Yeah." Rhona looked down.

Contessa tilted her head, a slight scowl forming on her forehead. "What's wrong with that? What aren't you telling me?"

Rhona moved her head up briefly then directed her focus back to the floor. "She wants me to go with."

"That's fine. We can go, and I'll show them where I fell into the river."

"No!" Rhona jerked her head up. "You need to stay here and get better.

I'll show her where we went swimming and where we found you. That should be enough."

Contessa felt helpless, something she wasn't used to. Raising her voice, she asked, "So I should just sit here all day? Doing nothing?"

"Getting well. That's not doing nothing."

It certainly seemed like nothing. Her own mother wouldn't let her help. She recruited Rhona when she got too tired. She smiled thinking of Rhona's reaction to being cooped up in the house. So eager to go out and do something. Maybe that was why she'd agreed to let Contessa walk outside and sync her log entry to the Oracle.

"You need to take another drink," Rhona prompted and waited for another swallow. "We're leaving in the morning, the day after tomorrow."

"It will be a dark day."

"Yeah. Kellach says he'll pack the rover with lights." She changed subjects. "Did you hear that Mrs. Stanton is pregnant? It will be the first baby on Niton." Rhona prattled about how a couple had brought their two dogs to Niton and how she just learned about it. "I didn't know we could bring pets!" Contessa smiled because she had already known about the dogs. Rhona went on about other things happening with the colony. Little inconsequential things, but the way she shared them helped Contessa take her mind off herself.

Hello, future historians! Rhona reporting in. My last trek was horribly interrupted, but today we have the master mechanic himself with us so if we run into any problems they can be fixed. I'm going with Dr. Yamamoto and Kellach to find out what made Contessa sick. Maybe today I'll be the one to find a new animal species. (System mark: Day 33)

Rhona shivered as she strode to the maintenance building. Only the slim crescent light of Hoyle lit the dark day.

"It is cold today." Kellach's voice startled Rhona. "Me? I'd be wearing a jacket."

Rhona jerked around. *That's creepy. How long has he been following me?* "Hi, Kellach. Yeah, I was just thinking I should go back and get my coat."

"Probably for the best. It's only going to get colder."

"Okay. I'll be right back. Don't leave without me." She briskly walked back home. When she reached for the front door, it opened ahead of her to reveal Contessa in a coat.

Contessa gasped, visibly surprised by Rhona's presence. "You scared me. What are you doing here?"

"Me? What are you doing with your coat on?" Rhona could barely keep the irritation out of her voice.

"I'm going with."

"Oh, no, you're not! You have to stay here. What if we find more of whatever made you sick?"

"Then I'll be immune because it's already in my body."

Rhona's mouth dropped open. *Does she think I'm stupid?* "That's not how toxins work, and you know it!"

"I do, but I didn't think you did. I want to go. I can't just sit here all day."

"We've been over this." Exasperated, Rhona marched past Contessa to their bedroom. She pulled her jacket out of the closet and put it on. She faced Contessa. "You can't come. You want me to get Dad?"

"He left long before you did."

*Of course he did.* As an early riser, it only made sense he'd already left. This was Rhona's chance to do something on her own. A chance to prove herself. She didn't need her big sister watching over her. "Then I'll get mom. I'll get Dr. Huff! You can't come!" Her voice rose in volume as she talked.

"I can so! I'm fine!" her sister yelled back.

"I don't need you!" Rhona immediately regretted saying it. She might as well have asked her sister to roll over and die. The pained look on Contessa's face made her feel worse. "I mean I don't need you on this trip. I'll be with Dr. Yamamoto and Kellach. I'll be fine."

Contessa took her coat off and collapsed onto the bed with a deep sigh. "Contessa—"

"No. You made it clear. Go on your grand adventure. I'll just sit here." "That didn't come out right." Rhona sat next to her sister. "I'll always need you. This is just something I want to do myself. I'm sure you can understand that."

Contessa sat there, not saying anything.

Rhona put her arm around Contessa's shoulders. "You can understand, can't you? A time to prove myself."

Contessa held her gaze. "Promise me you'll be safe?"

Rhona nodded and wiped her eyes. *Stupid tears.* Contessa put her arms around Rhona's waist, and Rhona hugged her back for several seconds before letting go. "They're waiting for me."

"I know." Contessa stood. "I want a full report when you get back."

"Yes, ma'am." Rhona jumped up and left the house. *This is going to be so much fun.*

Dr. Yamamoto waited with Kellach. She greeted Rhona with a smile. "All set?"

Rhona nodded. She'd been born ready.

"Great. I like the longer days when it's light out, not so much when it's dark. Seems no matter what I do, I can't get my body to live beyond a twenty-four-hour day." Dr. Yamamoto got into the front passenger seat of the rover.

"I once heard our bodies run on a twenty-five-hour clock." Rhona climbed into the front seat next to Dr. Yamamoto.

"That's an old theory, but other studies have disproven it. Seems we're genetically wired for a twenty-four-hour day. An Earth-based circadian clock, if you will."

Kellach drove through the bay door and tapped the buttons on the console to have the rover drive itself to the waterfall. The ride somehow went by more quickly this time, even though it was going to be a few hours before they reached their destination.

Why couldn't they make these rovers go faster? They only went up to sixty kilometers per hour. When she'd asked about it on her first trip with Fanar and Oswin, Fanar had droned on about battery chemistry favoring duration over peak current. He said more, but she stopped listening at the mention of chemistry.

The darkness outside changed the landscape to shades of grey. This, combined with the constant whine of the motor and the motion of the vehicle, made it easy for Rhona to slip into a light sleep.

"Okay, here we are." Kellach's voice woke Rhona up. She yawned and peered out the window. She could never let Contessa know she fell asleep. Koch had risen, giving the cascading water beads the appearance of falling pearls.

"Oh, it's beautiful." Dr. Yamamoto climbed out, letting in the full volume of the falls. She closed her eyes as the mist covered her face. She opened her eyes and watched the water for nearly a minute before wiping her face and climbing in so she could be heard. "I'll have to come back when it's daylight. So, where did Contessa fall in, and where did you find her?"

"First we drove down to get away from the loud noise." "Okay, let's do that, then."

"Okay," Kellach said. "This is where the extra lights will be helpful." He tapped a button on the console, and lights from the rover brightly illuminated the area. Several little creatures ran to escape into the deep shadows. "Just say when."

They drove forward until Rhona gave the word to stop. She climbed out and looked at the river. Yes, this appeared to be the spot.

"That's an old theory, but other studies have disproven it. Seems we're genetically wired for a twenty-four-hour day. An Earth-based circadian clock, if you will."

Kellach drove through the bay door and tapped the buttons on the console to have the rover drive itself to the waterfall. The ride somehow went by more quickly this time, even though it was going to be a few hours before they reached their destination.

Why couldn't they make these rovers go faster? They only went up to sixty kilometers per hour. When she'd asked about it on her first trip with Fanar and Oswin, Fanar had droned on about battery chemistry favoring duration over peak current. He said more, but she stopped listening at the mention of chemistry.

The darkness outside changed the landscape to shades of grey. This, combined with the constant whine of the motor and the motion of the vehicle, made it easy for Rhona to slip into a light sleep.

"Okay, here we are." Kellach's voice woke Rhona up. She yawned and peered out the window. She could never let Contessa know she fell asleep. Koch had risen, giving the cascading water beads the appearance of falling pearls.

"Oh, it's beautiful." Dr. Yamamoto climbed out, letting in the full volume of the falls. She closed her eyes as the mist covered her face. She opened her eyes and watched the water for nearly a minute before wiping her face and climbing in so she could be heard. "I'll have to come back when it's daylight. So, where did Contessa fall in, and where did you find her?"

"First we drove down to get away from the loud noise." "Okay, let's do that, then."

"Okay," Kellach said. "This is where the extra lights will be helpful." He tapped a button on the console, and lights from the rover brightly illuminated the area. Several little creatures ran to escape into the deep shadows. "Just say when."

They drove forward until Rhona gave the word to stop. She climbed out and looked at the river. Yes, this appeared to be the spot.

"I can so! I'm fine!" her sister yelled back.

"I don't need you!" Rhona immediately regretted saying it. She might as well have asked her sister to roll over and die. The pained look on Contessa's face made her feel worse. "I mean I don't need you on this trip. I'll be with Dr. Yamamoto and Kellach. I'll be fine."

Contessa took her coat off and collapsed onto the bed with a deep sigh. "Contessa—"

"No. You made it clear. Go on your grand adventure. I'll just sit here." "That didn't come out right." Rhona sat next to her sister. "I'll always need you. This is just something I want to do myself. I'm sure you can understand that."

Contessa sat there, not saying anything.

Rhona put her arm around Contessa's shoulders. "You can understand, can't you? A time to prove myself."

Contessa held her gaze. "Promise me you'll be safe?"

Rhona nodded and wiped her eyes. *Stupid tears.* Contessa put her arms around Rhona's waist, and Rhona hugged her back for several seconds before letting go. "They're waiting for me."

"I know." Contessa stood. "I want a full report when you get back."

"Yes, ma'am." Rhona jumped up and left the house. *This is going to be so much fun.*

Dr. Yamamoto waited with Kellach. She greeted Rhona with a smile. "All set?"

Rhona nodded. She'd been born ready.

"Great. I like the longer days when it's light out, not so much when it's dark. Seems no matter what I do, I can't get my body to live beyond a twenty-four-hour day." Dr. Yamamoto got into the front passenger seat of the rover.

"I once heard our bodies run on a twenty-five-hour clock." Rhona climbed into the front seat next to Dr. Yamamoto.

Kellach got out and inspected the ground. "Can't tell with this ground cover. No tire tracks."

"This is where we went swimming. Contessa doesn't like to swim, so she, ah, went for a walk along the river."

"We can't drive along the river directly. Too many bushes and trees," Kellach said.

"Yeah, we had to go around when we searched for her." "Do you know how far you drove?"

"Not exactly, but I remember seeing a big brown boulder on the other side. I think it was about two kilometers."

"Really?" Dr. Yamamoto turned. "I wonder what made it brown."

Rhona shrugged. "The ground was all brown too. Like a thin brown suede."

"Was it slippery?" "No."

"Interesting. I'll be right back." Dr. Yamamoto grabbed the bag at her feet and got out of the vehicle to get a sample from the river. She also took some soil and plant samples from the area before climbing back in.

Kellach started the rover moving. "Okay, so keep an eye out for that boulder across the river."

Rhona peered into the darkness. The bright lights were helpful, but the way the shadows from the trees moved caused confusion. Something would seem to be there, and then it would be gone as the area was illuminated.

When they got close to the two kilometers, Kellach brought the vehicle to a slow crawl. Rhona rolled her window down in an effort to see more clearly. She pointed and opened her mouth to say something, but then dropped her hand. Whatever she'd thought she'd seen had disappeared with the shadows.

"We've gone three kilometers." Kellach stopped the rover. "What?" Rhona scowled and looked at him. "Are you sure?"

He nodded toward the console. "Uh, yeah." "Was someone out here? Maybe they moved it."

"Nobody was scheduled to come out here, and they definitely wouldn't be moving a boulder," Dr. Yamamoto said.

Rhona's eyes went wide. "Then what happened to the boulder?"

# CHAPTER TWELVE

## The Cave

Kellach turned the vehicle around. "Let's take a different approach. Watch the river and see if you can identify where you crossed."

Rhona nodded. "Okay, hold on." She got out, entered the back seat, and sat on the driver side with the window down. The moving shadows still posed a problem as they crept along. She was wondering if she would ever find the spot when she recognized the alignment of the trees. "Stop!"

Kellach halted the vehicle.

She climbed out of the rover, walked to the river's edge, and focused on the other side. Yes, this was definitely the spot. She got back inside. "This is where we crossed."

"You're sure?"

"Definitely. I know because of those two trees that form a 'V' shape." Kellach nodded and inched into the river.

"This is scary. What if the river is too deep?" Dr. Yamamoto peered out the window with wide eyes.

He chuckled and continued forward. "This is truly an all-terrain vehicle. It floats and has an aqua propulsion drive."

"I can see the value of using this for all the off-world travels."

After they crossed, Rhona scrambled out, ran a short distance, and stopped. She spun about and looked. This was definitely the spot, but she didn't see a depression in the ground or drag marks anywhere indicating the removal of the boulder.

"This is an unusual substance." Dr. Yamamoto squatted and ran her gloved hand across the surface. "It's similar to moss, but seems to be woody like cork." She took a sample of it and some samples of other plants nearby. She inspected the general area. "It's covering a significant area." She sniffed and frowned before standing up. "I think we need to move away from here."

Rhona sniffed too. "I don't smell anything."

Kellach did the same. "Me neither."

Dr. Yamamoto approached the rover. "That's what bothers me. I have a keen sense of smell, and I'm not smelling anything at all. Like something is killing my ability to smell."

Rhona held her breath and ran to the rover. Once inside, she rolled up her window, hoping the rover had good air filters.

Dr. Yamamoto smiled as she got inside. "I don't think we're in danger. Just a precaution. After all, you were here once before, no?"

Kellach looked up at the trees and smiled before getting into the driver seat. "So you think this ground covering might be what's making the young lady sick?"

"We won't know until we do an analysis. I must confess, my gut is telling me there's something here."

Kellach nodded and tapped the console to mark the place on the map. "Okay, is that everything we need?"

"I believe so. Water and plant samples from the area will hopefully give us the information we need."

Kellach killed most of the external lights and maneuvered the rover back toward the waterfall.

Rhona could stare at it for hours. The moonlight reflecting off the water in front of the dark—wait, had she just seen a light from behind the water? "Stop!"

Kellach brought the vehicle to a standstill. "What's wrong?"

Rhona got out and stared. It appeared normal. Pearly white beads of water in front of a dark cliff thundered onto the submersed rocks, creating a perpetual fog just above the riverhead. She got back inside. "I thought I saw something behind the water. Can you turn on the lights again?" She waited for the lights, but it wasn't any better. All that water just reflected the light. "Is this thing able to go into the waterfall itself?"

"No. It can float, but it isn't designed to take a pounding like that."

"Is there a way around? Maybe there's a way to get behind—" A thump on the side of the vehicle startled her before the door opened. She jerked back and screamed.

Fanar and Oswin clambered inside and closed the door. "Are we glad to see you!" Oswin exclaimed.

Rhona stopped screaming. *Aren't they on Earth?* "Fanar? Oswin? How?"

Oswin caught his breath. "If the lights hadn't been on, we would've missed you."

Rhona could hardly believe her eyes. "Where did you come from?"

Kellach had shifted in his seat. "I thought you stayed on Earth. How did you get way down here?"

Fanar answered in short bursts between breaths. "I have no idea. We had to leave early because some Greeners damaged the power reserves and the technicians didn't know how much time was left on the portal."

Rhona listened, still finding it hard to believe they were right in front of her.

"Did you hear that pop?" Oswin's eyes went wide. "I thought we were goners."

Fanar looked at him. "I did!" He continued his narration. "We barely made it through, and pop! That's the last thing I remember before Oswin woke me up."

"That was two days ago!" Rhona exclaimed. "What happened? How did you get here?" Rhona waved her hands in the air as she spoke.

Fanar's jaw dropped. "Two days!?"

"How's that possible? We just woke up," Oswin said.

"It was, what, maybe twenty minutes ago Oswin woke me up." Fanar glanced at Oswin for confirmation. "We heard the waterfall and walked toward it. Figured we could find our way to Landing from there."

Oswin smiled. "When we saw the lights, we could hardly believe our eyes. We started running. Well, I started running. Not sure what you would call that thing Fanar did."

Fanar shook his head. "Hey, I'm not the track star, but I did okay." "Actually, you did. Your endurance is getting better."

"We should get you two back so the doc can take a look at you," Kellach said and put the vehicle into motion. "So the facility was attacked?"

"Yeah, the police had to escort us inside," Fanar said.

Kellach glanced back. "That popping noise must have been an explosion. Probably caused a surge that temporarily moved the target just before it closed. That would explain how you got down here." Fanar agreed. "That makes sense. Did Randy make it back?"

Rhona nodded. "They were mad when he returned without you. He kept saying you were right behind him."

Fanar and Oswin replied in unison, "We were!"

"We also found what made Contessa sick. It's that plant she was lying on. It kills your sense of smell and then kills you."

Dr. Yamamoto chuckled. "We're not sure about that. Dr. Huff and I need to do an analysis of the various samples."

Fanar sighed. "So what made it through the portal?"

"The portal generator. That and another platform of supplies that came through before its premature closing. I'll need help installing all the items," Kellach said.

Rhona hugged Fanar and awkwardly tried to hug Oswin too. His position on the other side of Fanar made it difficult. It was good to see them. "Well, I'm glad you're back. It wouldn't be the same without you." She watched the waterfall disappear behind them. She was certain something was there and wondered how long it would be before she got a chance to investigate it.

This is Fanar. I'm back and helping with the assembly of the return machine. It's been a week and we're almost done. Because the portal closed early, we didn't get all the equipment we had needed for repairs. Contessa has been quick in recovery, and Rhona has been asking Oswin and me about the possibility of going back to the waterfall. She swears she saw something that is worth investigating. (System mark: Day 41)

The sun slowly sank into the distant western horizon. Fanar held a solar panel up to the disappearing light, observing how the dark surface showed tiny rainbows of color when observed at certain angles. It reminded him of the portal. He handed it to Kellach's waiting hands. "These panels are lighter than what I'm used to."

Kellach mounted the panel onto its bracket. "They're the latest available. Up to eighty percent efficiency."

"Wow. Will it store up power for the portal faster, then?"

"If we had a larger field of them, yes. But we only have a couple dozen, so it will still take a month to get a full charge." Kellach connected the power line. "That should do it. I have to coordinate some other items." He glanced at the setting sun. "Why don't you take the afternoon off?"

"Thanks. Rhona's been bothering Oswin and me about taking another trip to the falls. I think we should make it an overnight trip."

"That's a good idea. The next two days will be sunny days. We have some useful gear in storage room five. Take what you need and make sure you reserve a rover for the next two days."

Fanar nodded. "I will." He ambled across the field—which he called 'The Circle'—in the middle of the main buildings and down the road to the building Oswin worked in with his parents. Inside, metallic counters and cabinets lined the outer walls. Down the center ran a long table with two empty aquariums. Oswin stood next to the repaired microscope, peering into the eyepiece. His parents and a few others were nearby at different stations.

Fanar tapped Oswin on the shoulder.

Oswin jerked back in surprise. "Fanar! You scared me. Aren't you supposed to be helping Kellach?"

"He gave me the rest of the day off. You know how Rhona wants to explore the falls again?"

Oswin rolled his eyes. "I'm surprised you weren't accosted on your way here. She and her sister just left a few minutes ago asking if I'm busy the next two days."

"I was just talking to Kellach about it. He said there's some overnight gear we could use."

Oswin put his sample into its storage case and placed it into a drawer. "A camping trip? We'll have to plan it out properly."

"I guess so." The last time he'd been camping had been with his mom several years ago. She was a nature lover and would take him and Oswin on hikes and, occasionally, a camping trip. His dad would go along if there were good opportunities for some stargazing.

"Hey, Mom? Think you can get along without me for a couple of days?" Oswin asked.

She winked and smiled at Fanar. "Somehow we'll manage. But we'll definitely have some new things for you when you get back and expect you to be focused."

"Okay." He said to Fanar, "Let's see what gear's available."

The road panels glowed as the sun disappeared behind the mountains on the far side of the lake. They went to the maintenance building, and Fanar led the way through a back door to the storage room Kellach had talked about earlier. Inside, they found small lanterns, a few tents, empty containers for water, and sleeping gear. Oswin counted off with his fingers as he went over the items needed.

Fanar rummaged through the containers but didn't see anything to eat. "Let me check next door. I think there's something in there to eat." He walked around to the neighboring room and found the door locked. He scowled and muttered to himself, "Hmmm. I don't remember that door being locked before."

"We can find something from home. It's only two days." Oswin piled the items into a container. "Which rover we taking?"

"Oh, yeah. Let me go see which one is available." Fanar led Oswin through the building to the front room. Along the way, he searched in other rooms for food that could be taken. Two other doors were locked. "I don't understand why those doors are locked."

"I know we're locking the lab. I've heard other buildings are being locked too. It's a shame that we can't trust each other."

"Who do you think caused all that damage?"

Oswin shrugged. "I have no idea. Everyone I talk to seems as baffled as we are. It's like there's an invisible person living with us."

Fanar reserved a rover for the next two days and committed the access code to memory. They returned to the rover and loaded their items into the rear segment before walking to their house.

A pounding on the house door penetrated Fanar's slumber. He could hear his dad talking and the unmistakable voice of Rhona. She was definitely eager to do this.

"She's here already?" Oswin groaned.

"Don't blame me. She's your girlfriend."

The ceiling glowed in time for Fanar to see an airborne pillow about to land on his face.

"She's not my girlfriend. You can have her," Oswin said.

Fanar scoffed and threw the pillow back. "Like I could handle all that energy."

A light rap on the bedroom door was followed by his dad's voice.

"Fanar? Oswin? You guys awake?"

"Yeah, Dad. We'll be right out." Fanar sat up and stretched. "Guess we should get going."

After getting dressed, they met the girls, who sat on the couch in the living room. Rhona stood. "Okay. Let's go."

Fanar held his hand up. "I gotta get something to eat first." Contessa smirked. "I told you."

Rhona frowned and plopped back down. After staring at Fanar for a couple of seconds, she waved at the kitchen. "Well, hurry up!"

Fanar followed Oswin into the kitchen. He glanced back and leaned in to Oswin. "Yeah, you can keep her."

Oswin back slapped Fanar's arm and shook his head as he chuckled.

After eating, they gathered food they could consume without preparation.

When they got outside, Randy sat in a rover waiting for them. "Randy? What are you doing here?" Fanar asked.

"I volunteered. Mike and I need something to do, so we go with teams on their surveys and stuff. You guys are more fun; besides, you never know what you're going to run into." He patted the gun on the seat next to him.

"Who's Mike?" Rhona asked.

"My big brother. So glad he's here on Niton, 'cause the others don't include us in anything. It's like they don't trust us."

Fanar sighed and stowed the food in the back segment, noting all the gear they had already packed.

While the rover carried them to their destination, Fanar and Oswin gave a detailed account of their trip to Earth. Randy added to their account when he could.

Contessa's eyes went wide. "What are the Greeners so upset about?"

"It's obvious," Fanar said. "Wherever we go, we junk it up. First Earth, then the moon, Mars, and now Niton. We need to learn how to live in a way that doesn't leave a bunch of junk and pollution behind so we don't mess up any other worlds we travel to." He was opening his mouth to say more when he realized he was spouting the Greener lines. Embarrassed, he kept his mouth shut.

"Well, I think it's exciting and they need to leave us alone," Rhona said. "I still think they sabotaged our equipment before we came over. The only thing that's happened since the first few days is the broken equipment in that lab."

Fanar nodded. "True. Maybe someone accidentally broke it and didn't want to admit it."

"Don't forget the missing dishes," Randy interjected. "I was almost kidnapped," Contessa reminded them.

That statement set them afresh on discussing whether a Greener was in their midst or not and whether anyone had been around Contessa. Maybe she'd gotten tangled in some sort of plant.

When they arrived at the falls, Rhona was the first out. She ran up to the water's edge and peered at the water pouring down. She pointed and yelled over the roar, "See how it's darker in the middle? I think there's a cave!"

Oswin wiped his eyes and examined the area. "I don't see it."

Fanar looked up past the mist at the base of the falls and scanned back and forth as he slowly lifted his gaze. Near the top of the mist, where it thinned, he made out a blurry dark spot. "I think I do. See where that rock sticks out about a quarter of the way down?"

Oswin scrutinized the area and eventually nodded. "About three meters to the right and straight down."

"I don't—oh, wait. I think I see it. Yeah, I guess it could be a cave. But how do we get there?"

Fanar walked toward the cliff and faced the falls. He shook his head and went back to the others. "Not from this side. But see how it's angled? I bet if we cross the river and come back up, we could maneuver behind the falls and maybe find a way to it."

# CHAPTER THIRTEEN

## Spelunking

**O**swin headed to the rover. "Let's get inside and talk. I'm tired of yelling."

The moment the last door closed, Rhona spoke up. "How are we going to get up there?"

Oswin looked at Fanar. "What do you think?"

What did he think? At first he thought this would be a waste of time. Caves were dark and damp and, if one were unlucky, housed dangerous animals. However, it could also be a place where one could shelter from the elements. What if his mom had been displaced when going through the portal like Oswin and him? Then a cave could prove useful. What if Rhona had caught sight of a light because of his mom? It could have been a fire that was momentarily visible. A renewed hope woke in him as he contemplated this. Water, food, and lodging. Basics that would allow someone to survive were all here. Maybe he could find his mom after all. He knew it was a long shot, but still in the realm of possible.

Fanar pointed to the top of the falls. "See how the left edge juts out? It creates a large area behind the falls on the other side. I guess we can see if it's possible to travel behind them from the other bank and maybe find a way up."

Randy followed Fanar's hand and nodded. "It might work."

"Thank you!" Rhona said. "You won't be disappointed. I just know there's something in there worth finding. When we were coming back from getting those samples, I saw a flash of light coming from the cave. So I know there's something there."

Another thought suddenly occurred to Fanar. "You saw a flash?" He looked at Oswin. "Yeah. Why?"

Oswin answered, "After we ran through the portal, I thought I saw a flash of light before passing out."

"I think it was just the sun," Fanar said. "Dr. Huff said we're lucky. Only dehydrated. There doesn't seem to be any long-term damage from our incident." He glanced at Oswin and chuckled. "Except for Oswin's hallucinations."

"I was not hallucinating!"

As Randy moved the vehicle forward, Fanar asked, "Did they find out what made you sick, Contessa?"

"Not yet. Dr. Huff said it may have been a combination of things. She said the Oracle wasn't finding anything from the samples that could make me sick."

"The Oracle doesn't know." Oswin grinned.

After traveling a short distance along the river away from the falls, Randy directed the rover toward the river and crossed. Then he maneuvered it through more trees, gradually working his way back to the falls. The trees thinned, making progress easier. When they got to the cliff face next to the falls, he stopped.

"Stormin! A path!" Rhona jumped out and ran to the front of the vehicle. Behind the falls, a flat surface about two meters wide was submerged just below the surface of the river and slowly rose up about five meters to the cave clearly visible with the entrance angled toward them.

Oswin smiled at Fanar. "Guess we're going spelunking."

The boys helped the girls with their packs before putting on the heavier packs.

Rhona grinned back at the others as she led the way. "Can you believe how lucky we are?"

Fanar ran his hand against the rock face. It was smooth from the constant exposure to the water. The path appeared worn and textured. Nothing rising or falling more than a couple centimeters, so they didn't have to worry about tripping over any rocks or pits.

His mom loved nature. She loved camping and hiking and learning about plants and animals. It wasn't just a job for her, but life itself. His dad loved the stars and all things floating in space. He could spend hours examining star charts and calculating the gravitational influence between celestial bodies or talking about the formation of the planets. Between them, they had the universe covered. On the trips with his mom, they had come across many waterfalls, and Fanar had never seen anything like this. Sure, there had been caves. Old rivers that had changed courses or been blocked by the build-up of minerals. But this path somehow seemed wrong. It looked normal, but something about it itched his subconscious.

The water fell out of reach to their right, but the mist enveloped them. The rush of the water left a cold breeze next to them. When they reached the cave, water dripped from their clothing. Fanar turned on his head lamp. The others followed suit. The cave was about three meters high and four meters wide and went straight back without narrowing. The light from their lamps faded into obscurity. The constant pounding of the falls was oddly muted.

"Looks like it gets dry ahead," Fanar said. "What do you think about finding a good place for the tents and changing into some dry clothes?"

Oswin nodded. "Sounds good."

They went deeper into the tunnel until the ground became dry. Fanar almost expected to see a place where a fire had been made. Proof of his mom's presence. However, no such proof could be found. He set his pack down and pulled out the two-man tent.

"We can help." Contessa's voice echoed along the tunnel. She put her pack down.

"No need. We got this," Oswin replied.

With practiced ease, Fanar set up a tent for the girls while Oswin and Randy set up the other tents.

Contessa walked over to Fanar and watched as he finished. "You make it look easy."

"We've done this a hundred times." He unzipped the opening to her tent.

"Here you go."

She smiled at him. "Thanks." She and Rhona went inside and disappeared behind the opening as she zipped it closed.

"How far back do you think it goes?" Oswin entered their tent and stripped off his outer layer.

Fanar zipped the opening closed and undressed. "I don't know. Pretty amazing about the path being there."

"Yeah, we lucked out." Oswin put on his shirt. "Warm in here too. Think there's a hot spring somewhere nearby?"

"Wouldn't that make the lake warmer too?" Fanar asked. "Though that brings up a point. This is probably a lava tube, not the remnants of a river." Oswin unzipped the opening and stepped out. "Why do you say that?"

Fanar joined him and pointed at the wall and ceiling. "It's too smooth to be from a river."

Randy was out of his tent, watching and listening as Fanar explained. After Fanar finished speaking, Randy said, "You really know a lot, don't you?"

"He's the smartest person I know," Oswin said.

Fanar stood a little taller, basking in the praise.

"Are we safe in here?" Contessa spoke from inside her tent. "I'm sure we're fine," Rhona said. "Hurry up so we can go."

Oswin smiled at Fanar. His headlamp dimmed automatically.

"Someone's in a hurry."

Rhona came out of her tent, followed by Contessa. "How much time do we have?"

Fanar tapped his uplink. No response. He frowned and tapped again and was able to see the time. "I figure we can go for about two hours before we need to turn back."

"Let's go!" Rhona stepped forward.

"Hold on." Oswin reached into the tent he and Fanar were using and pulled a smaller bag out of his hiking pack. "We packed these separately for spelunking. Water and waste management."

Rhona stopped. "Okay." She went to her pack and grabbed the small bag that resembled the one Oswin held. She scowled. "Um, waste management?"

Fanar chuckled. He couldn't stop. After several seconds of chuckling, it turned into laughter. Soon the others joined him. It was true. Laughter was contagious.

Oswin wiped his eyes. "That's funny. We should go." Randy took point as they headed down the tunnel.

Rhona ran up next to him. "I hope I don't have to pee."

They hiked through the tunnel for twenty minutes and stopped. In front of them, the tunnel opened to a large chamber.

Rhona swung her lamp from side to side. "Wow. Reminds me of the departure chamber on Earth. Only bigger."

Fanar nodded. "It must be massive. Even our voices are dulled. No more echo."

"Do you see the top?" Oswin directed his lamp upward. Fanar squinted. "I think I do?"

Rhona faced the others. "Let's go straight!"

For reasons he couldn't explain, Fanar hesitated. He pointed to the wall. "It's best if we keep this on our right. That way we don't get lost."

Her face brightened. "Oh! Good idea!" She turned and walked alongside the wall.

Randy caught up to Rhona. He looked back at the others, motioned with his thumb at Rhona, and smiled at her enthusiasm.

Inside, Fanar cursed himself. If he'd known about this, he would have brought the interior scanner. It housed thousands of tiny lasers to detect the size and shape of a large space and store it in memory. Perfect for mapping cave systems. Perhaps he hadn't really believed there was anything here— perhaps he had thought that Rhona had seen a reflection of moonlight.

They came to another opening and entered. This tunnel grew smaller and rougher. They traveled for twenty minutes following its winding path until they came to a fork. Rhona halted and faced the others. "I need you to stay here. I'll be right back."

"No!" Fanar sounded harsh even to himself. He lowered his voice. "We need to stay together."

She nodded. "I know. I just—" She sighed. "I, um, need to use the waste management."

"Oh. You ever use one of these before?" Oswin asked. Even in the low light, Rhona's embarrassment was evident. "I guess not." He took her bag and pulled out a flat package and explained how it worked.

"Okay. I'll just be down a ways."

"Maybe I should come with," Contessa said.

"Are you crazy? I can do this myself." She glanced at the item in her hand and trekked down the left-hand fork.

"I remember my first time using one of those." Oswin chuckled. "Wasn't fun with the solid waste."

"It's easier when you're out in the woods and you can just dig a hole," Randy said.

"Yeah, but some areas don't want you to do that either. Like the protected areas down the Colorado River."

"How you doing, Rhona?" Contessa called out. "Okay." Rhona's voice came from a distance. "How far did she go?" Fanar asked.

Oswin shrugged.

They waited a couple more minutes before Contessa called out to her sister again.

"This isn't funny guys. Where did you go?" Rhona asked.

"We haven't moved. Just come back the way you went." Contessa asked Randy, "Should we go after her?"

Randy shook his head. "No. We need to stay in one place so she can find us. She can't be too far."

Contessa stepped toward the left fork. "Just follow my voice. Come back this way. Remember, you went left, so now you want to come right."

"Left?" Rhona's voice was farther away.

Contessa raised her voice. "No! Come right!" She searched for encouragement from the others and took a deep breath. "Rhona?"

"I can't find you." They could barely hear her thin voice.

Contessa turned back. In a rush she asked, "What do we do?" She looked down the tunnel and then back. "We have to do something!"

Fanar said to Oswin, "Looks like we have no choice. Do you have some cord?"

"I might." Oswin hunted in his bag, closed his eyes, and sighed. "If I did, it's back with the tents." "Great."

Randy suggested, "Maybe if we all yell together, we can get Rhona to hear us."

Contessa was visibly encouraged. "Yeah. Let's try that."

At the count of three, they yelled out Rhona's name then listened. Fanar asked the others, "Did you hear her?"

"I thought I heard something," Contessa said. "Let's try again." "Rhoooonaaaaa!" They yelled her name, the sound of their voices echoing down the tunnels.

They listened for a response.

"Okay. Let me start recording our movements." Fanar tapped his uplink. Nothing happened. He scowled and tapped it again without a response. He looked up. "Mine isn't working. Can one of you use yours?"

"Are you sure?" Oswin asked. "Those things never fail."

Fanar tapped it again. "See? Nothing. I must have damaged it somehow."

The others exchanged glances. "I think you're the only one who brought one, Fanar," Oswin said. "I haven't used mine in so long I don't even think about it anymore. Without the linknet, it's almost useless."

Fanar was appalled. Was he the only one who had any appreciation for technology? "That's… inadequate!" He huffed and peered down the tunnel that Rhona had taken. Even with Randy there, he felt responsible for their wellbeing, even if his dad didn't ask him to be.

Oswin stifled a laugh. "If we all work together, I'm sure we'll be fine. It can't be that complicated. She obviously got her left and right confused."

"She always had a hard time with that," Contessa admitted.

Randy took charge. "Okay. We need to stick together. No more separating." He led the others down the tunnel. The rough grey stone surrounded them as they stepped through a narrow spot and continued.

After another thirty meters, another fork appeared. They called out for Rhona down each fork and waited for a response.

Fanar asked Contessa, "Which way would she have gone?"

Her eyes were wet with unshed tears. She shrugged. "I don't know." She looked back the way they'd come. "Maybe right. Because if she accidentally came this way instead of going back that way, then right would have been… right."

Fanar nodded. "But then she called back about going left." "We gotta pick a way. I think right is best," Oswin said.

"Okay. Right it is." Randy led the way down the right tunnel.

The tunnel bent and sloped up. After several minutes of walking, they called out to Rhona again.

Oswin pulled up. "I'm not sure this is the way. Let's try the other way."

The others agreed and went back the way they'd come. To their left, by a boulder, stood another tunnel.

*Where did that tunnel come from?* Fanar was sure it hadn't been there a moment ago. With that boulder there, it was possible he had missed it in the shadows. He was becoming unsure of himself. He couldn't even remember that boulder being there before.

"Left!" Contessa sounded relieved. "Remember when she thought she should go left? She must have been coming back and saw this tunnel!" Oswin shrugged. "Sounds good to me."

Randy nodded and led them into the new tunnel, his face tight with worry. They continued for nearly thirty minutes, calling for Rhona occasionally before they entered another large chamber, not nearly as substantial as the first one they'd encountered, with three other tunnels leading away from it.

Fanar thought fast. They were getting farther away from the place where they'd first lost her. If they didn't find her soon, they would be lost. He grew more frustrated with each passing minute. "Let's go to each of these tunnels and call out to her. Maybe she's down one of them."

Contessa wiped tears from her face and nodded.

They went to the first one on their left and stood at the entrance. After yelling out for Rhona and not getting a response, they tried the second tunnel.

Contessa gasped. "I heard something." They listened and didn't hear anything. "Let's try again. I know I heard something."

At the same time, they yelled Rhona's name as loud as they could and then listened. After a second, they heard her calling back. It was faint, but they definitely heard her respond. They yelled again and waited.

"Well, where is she?" Oswin asked.

"I'm here!" Rhona came up behind them.

"Rhona!" Contessa grabbed her sister in a fierce hug. "I thought I'd lost you!" She and Rhona both cried.

Rhona reached to hug Fanar then paused when she saw the angry expression on his face. "I'm sorry. I went the wrong way. I kept trying to find my way back but got more lost."

Oswin enfolded her in his arms. "At least we found you. You okay?" "Yeah." She hugged Randy.

Fanar put his hand to his forehead and rubbed it while he collected himself. "Where did you come from?"

Rhona pointed to one of the tunnels. "This place is like a maze. When I heard you calling me, I yelled out and started running."

Fanar frowned. "That's the tunnel we just came through."

Oswin indicated a different tunnel. "No, that's the one we came down." He looked at Fanar, his expression mirroring Fanar's own rising panic. "I think."

# CHAPTER FOURTEEN

## Stargazing

**H**ow difficult can this be? Fanar clenched his jaw. There were only a few forks; they shouldn't be lost so easily. Fanar ran into the tunnel he thought they'd come through.

"Fanar! Where you going?" Randy's voice chased after him.

"Nowhere!" he called back "I just have to see. I won't take any turns!" He ran for a couple of minutes, cursing himself. He should have handled this better. He should have planned better. He should have been... better. How was it he ran into things without planning? Not seeing any forks, he returned to the others. "Guess I was wrong." He put his hands on his knees and took in several deep breaths. "I was so sure we came down this tunnel."

"You okay?" Rhona asked.

"Yeah. Just not the runner." Fanar caught his breath and stood upright.

Oswin led them down the tunnel to the right of the one Fanar had just taken. They walked for several minutes and came to a fork. He stopped and looked both ways.

"We go left." Fanar pointed.

Oswin faced Fanar. "But there isn't a boulder here."

Fanar swung his head to the area the boulder should be, illuminating the spot with his headlamp. As Oswin said, no boulder. Not just that—the quality of the air was different. Fresher.

"This is where I was a minute ago," Rhona said.

Fanar put his hand over his face and rubbed his temples. *This is a disaster.* "Okay." He brought his hand down. "Let's try this again. I have an idea." He led them back to the large chamber they had just left, took a water bottle out of his pack, and set it next to the tunnel. *I should have done something like this earlier.*

Randy nodded, the light from his headlamp bobbing up and down the cavern wall. "Good idea. Let's try the tunnel to the right."

They proceeded for ten minutes before Fanar paused. "This can't be the right way."

"Listen," Oswin said. "I think I hear the waterfall."

No one breathed as ears strained to catch what Oswin heard. Faintly, Fanar made out the distant pounding of the water.

"Let's keep going," Contessa said. "Maybe we can get there from here." "I'm willing to try." Fanar moved forward. The tunnel gradually curved toward the right then went left, and after another fifteen minutes of hiking, they entered the original large chamber with several connecting tunnels.

"Finally! Something I recognize." Rhona ran into the large space. "This is the first one we came to, right?"

"I can hear the waterfall." Contessa shined her light into a tunnel just to the left of where they'd emerged from. "Down here; this must be the original tunnel."

Fanar looked at the tunnel. "I guess. The way these tunnels twist, I'll believe anything right now."

Contessa led the way down the tunnel, her steps getting faster as they grew closer to the sound of the falling water. Her lamp lit up the tents. She ran toward to them. "I never thought I would be so happy to see some tents!" "I am so tired," Rhona said as she approached their tent. "How long were we gone?"

Fanar shrugged. "My uplink stopped working, but I would say several hours. See? It's dark outside." He plodded to his tent. The flap was open. "Oswin, you forgot to zip it up." He crawled inside.

"You were the last one out." Oswin stood at the entrance of the tent.

"Anyone want to eat before we turn in?"

Fanar lay on his sleeping mat and closed his eyes. His body melted into the mat. "Sure. Let me just take a nap first."

*Thump thump.* Fanar opened his eyes. *Did I hear something?* It was soft but sounded close. He must have slept through the night. It wasn't bright yet, but the early rays of light entered the cave. Oswin's heavy breathing indicated he was still sleeping. *Thump.* Fanar heard it again, but farther away and barely heard between Rhona's snores. He snuck out of the tent. All he could see were the three tents. He crinkled his nose. That odor had returned. That decaying plant odor from one of the tunnels that wafted around. He approached the cave opening behind the waterfall and stopped at the edge of the dry ground. A million tiny rainbows glittered the cave walls as the water refracted the sunlight.

Fanar swiveled around and saw Oswin and Randy standing between the tents, the girls coming out of theirs. He walked toward them. "Did you guys hear that?"

"I just heard you rustling around in our tent and leaving," Oswin said.

"I heard you getting up," Contessa said then pointed at her sister, "and got her up."

Randy shook his head. "Sorry, didn't hear anything."

Fanar was the only one who'd heard it. Maybe there was something special about this place as Rhona had said, but he wasn't prepared for an extensive exploration. "We need to get back."

Rhona gazed down the tunnel and sighed before returning her attention to the group. She grinned. "It's not as exciting as what I'd thought, but at least I found something interesting."

Contessa nodded. "I'm sure you'll want to come back."

"You bet I will! Next time we'll bring equipment so I don't get lost. And more food!"

"Speaking of food…" Oswin reached inside his tent and pulled out a container. He opened it and passed breakfast bars to everyone. "I heard Dr. Yamamoto say that we'll need to start eating some of the native food since we didn't get back the expected supplies from Earth."

Fanar chuckled. "Will everything taste like chicken?"

Rhona laughed. "I hope not. I hate chicken."

They finished eating and packed everything into the hiking packs. Fanar looked down the tunnel before they left. His eyes darted from shadow to shadow, but only stone walls stared back at him. There was no evidence his mom had stayed here. Whatever he'd heard was gone.

"Sounds intense," Kellach said.

"I'm amazed we found our way out." Fanar closed the door of the rover he was working on. This one had been used by the team at the mining site, and it needed the air filters cleaned and depolarized. Kellach had assigned him to the task while he worked on something else nearby.

"Sounds like you had quite the adventure," Kellach said. "Yeah, I guess you could say that."

"You don't sound too excited about it."

Fanar nodded. "I used to go camping with my mom and dad, and it was fun. But I also enjoyed getting back to civilization. Here, I feel like I'm always camping."

"Me? I would hardly call this camping. Nice housing, working facilities, food—"

"True. Maybe camping isn't the word, but I do miss proper civilization and having technology that works."

"Ah, you mean the linknet. It would be more convenient if we had that."

"I'm not so sure about the food here either. I hear we'll need to start eating the local food."

Kellach grunted. "You know it's safe, right? Samples were taken by the drones and tested on Earth. Probably healthier for you."

"Yeah."

"Tomorrow we can start a full diagnosis on all the buildings to make sure nothing else is wrong and determine what can be replaced with the items sent from Earth. You about done there?"

Fanar held up the filter in his hand and nodded. "Let me just put it back." He lay on the floor of the rover and placed the filter into its housing under the seats. He wiggled his way out and stood. "Okay, what's next?"

Kellach looked around then back at Fanar. "That's it for today. Nothing I can't handle myself."

"Has the Oracle come up with anything about what made Contessa sick?"

"The Oracle? Oh, the IQ system. I've been working with the doctor, and it hasn't come up with anything. Me? I have work to do, so go do whatever it is seventeen-year-olds do these days." He left the bay and unlocked a room down the hall before entering. The distinct click of the door lock reverberated off the walls.

Fanar shook his head. Why couldn't he help Kellach? Another case of that man being a little off. He peered out into the dark day. Three more weeks and Earth would be opening the portal again for the last time. He smiled. Civilization, a normal day cycle, and technology that worked. *It must be nice.* Kellach seemed to minimize the technology. Fanar knew the golden age of electronics had passed. A time when even adhesive bandages had circuitry to monitor the wound. But he preferred what Earth had compared to what was available on this planet. He scrutinized the empty place on his wrist. The uplink had very advanced circuitry—not something that could be easily repaired. A new uplink would be the first thing he got from Earth given the opportunity.

He looked down the hall where Kellach had disappeared and grimaced. *Guess it's a nice 'day' for a walk.* He grabbed a hand light from the wall by the door and left. The glowing road panels were soon behind him as he made his way toward the lake. It took a little over an hour to get there, but he enjoyed the time by himself. The light exposed a few dips and small holes in the ground, allowing him to avoid a twisted ankle.

Fanar lay on the soft ground by the lake, turned off the light, and stared into the sky. The stars were so bright here without the light pollution from the cities on Earth. It was beautiful, and he enjoyed gazing at them, but it wasn't his sky. None of the familiar constellations could be found. He sighed again.

"Those three stars could be Orion's Belt," his dad said from behind him. The shadow of an arm pointed to the left.

Fanar jerked up into a sitting position. "Dad? I heard you would be away for a couple of days." He could just make out his father's figure.

His dad chuckled. "Sorry if I scared you." He pointed to the water pump several meters away. "I got back early and was checking on the equipment when I saw you come down." He looked up. "So beautiful, aren't they?"

Fanar could almost swear the stars moved. *Am I sitting?* Yes, he was still sitting. *He's talking to me!* It had been so long since he'd heard so many words from his father. He realized he hadn't responded. "Yeah. I miss the constellations you taught me on our campouts. You really think that's Orion's Belt? The other stars aren't there."

His father sat next to Fanar and leaned back, bracing himself against the ground with his arms. "You're thinking two dimensionally. If you think of the universe as a sphere and rotate it, then the star patterns appear completely different." "Oh."

They sat in silence, drinking in the display of creation. After a minute, his father spoke again. "I should have been with her." "Where?"

"Here. On Niton. When she went through the wormhole, I should have been with her." He focused on Fanar. "Did you know your mom named the planet?"

"No." Fanar held his breath. His dad hadn't said more than a few words to him since the day his mom had died. Disappeared. Fanar still couldn't make up his mind.

His dad stared at the lake. "I was just so sure of the technology—" He took a shuddering breath and paused for several seconds. "This was our thing. To find and explore a new planet, and she isn't here." He looked at Fanar again. "The project almost died with her. Well, my participation, anyway. I went into a depression."

Fanar remembered. For nearly six months, his dad had slept most days and hadn't talked to anyone.

"I found a plant sample from Niton in the bedroom by her necklace. She must have brought it from the institute. Why, I'll never know since it was against protocol. Then I decided. She deserved to be on Niton. Obviously, I couldn't bring her, but I could bring her necklace. Her favorite necklace." He paused as if considering whether to say what came out next. "I may have cut some corners in my sudden desire to be here." He wiped his eyes and studied Fanar. "You take after her, you know, in your appearance. Every time I see you, I think of her. I guess that's why I've been avoiding you."

The sudden reversal of his dad's isolation now made sense. Fanar didn't know what to say. How do you respond to such a statement? "It's okay."

"It's because of her you're here, isn't it?"

"Yeah." To say he hoped to find her alive sounded foolish now.

"You remember when we would go camping? You seemed happy to go, but when your mom or I would point things out, you would just acknowledge it. No questions, no enthusiasm. When it was time to go, you were the first to finish packing."

"I enjoyed being with you."

His dad smiled at him. "I guess I messed that up with all the time I spent on this project. Now look at you. All grown up and doing the responsible thing. You even came back when you had the chance to stay on Earth last week." He took in and released a deep breath. "But this isn't your dream. You don't need to stay here if you don't want to."

After this outburst of affection, Fanar believed his heart could burst. He'd come back to Niton because it was the right thing to do; however, now he had another reason. He might have lost his mom, but now his dad was here. Really here, and Fanar wanted to restore that bond.

Dr. Monier sniffed the air. "Do you know if someone brought some Stargazer lilies from Earth?"

Fanar sniffed. He could smell it again—like some old decaying vegetation. "Stargazer lilies?"

"They were your mom's favorite. Large pink lilies with white edges. White center. Spicy aroma."

"Yeah, I remember those. Very strong."

"When they get old, they smell awful. I would joke with your mom and say they matched her morning breath."

"Bad breath," Fanar whispered it to himself and went rigid. He scanned the surrounding area. *Is it possible?*

"You okay?" Dr. Monier asked.

Fanar jerked his head around. "Dad? Are we sure there isn't any intelligent life here? What if it's invisible?"

"Invisible? Why would you say that? No, there isn't anything here. We had drones flying all over and nothing in the pictures indicated a civilization of any kind."

"Contessa swore that someone with bad breath was at the river, but we never saw anyone. I also heard something in the cave and smelled the same odor there."

"Well, from what I gather, she was in pretty bad shape. She must have imagined it. As for the cave—"

Fanar jumped up to his feet. "I'm serious. I really think there's something here." In his mind, they were surrounded with some unknown visitors, like wolves circling a campfire. His heart raced. Were they being studied?

"Now calm down." Dr. Monier stood. "I'm sure there's a logical explanation. If there were intelligent life, some evidence like structures and remnants would be visible. And the cave and river are a significant ditance from Landing."

His dad made sense, but something didn't add up. Fanar smiled at his dad. "I'm sure you're right. I'm going to see if Oswin is done with his work."

"Okay. Maybe we can spend some time together."

"That would be nice." Fanar meant it too.

"Invisible?" Oswin stared at Fanar.

Fanar waited for a response. If Oswin didn't go for the idea, then maybe he had lost his mind.

Oswin repeated his question, but more to himself, and his gaze went through Fanar before refocusing. "Invisible actually makes sense. That's how they were able to damage our equipment without anyone knowing."

Fanar nodded. "I hadn't thought of that."

"It could be a natural thing. More like camouflage than actual invisibility." Oswin pulled his food tray out of the cooker and set it on the counter.

"My dad said if it's intelligent life, we should have seen some evidence like buildings."

"Why? Just because we like buildings doesn't mean all intelligent life does. In fact, I bet us putting buildings here is what got them upset!" Oswin's eyes grew big. "And that's why they sabotaged our stuff! To drive us away!" He paused. "You realize nobody would believe us."

Fanar frowned. "Yeah, I know."

# CHAPTER FIFTEEN

## Decisions

*Rhona reporting in. The portal opens in two weeks. Some people say Earth is opening the portal, and others say we're supposed to. Either way, it means I have until then to go exploring those caves again. Because once we get the supplies we didn't get last time, Fanar's going to be too busy repairing the ruined equipment. I get the strong impression Oswin won't go without him. Mama's been saying I need to find my own way to contribute to the colony. I have no idea what I can do here. (System mark: Day 47)*

"**W**hat do you mean, you're babysitting?" Rhona sat on her bed, watching Contessa put on a lightweight blue jacket. Since Dr. Huff had declared her healthy, Contessa had been anxious to be busy again.

"Small children need supervision when the parents take time out for themselves. It's called babysitting," Contessa said. Rhona made a rough noise in her throat. "I know what babysitting is."

"Then why'd you ask?"

"Why can't their teacher watch them?" Contessa frowned at her.

"I need your support. If you ask, then I'm sure he'll say yes."

Contessa stopped at the bedroom door. "Me? Why would he listen to me?"

"You're his favorite."

"I doubt that." Contessa rolled her eyes.

Rhona followed her sister to the house door. "Well, maybe so, but if you're with me when I ask, I bet he would say yes. You've been nicer to him than I have."

Contessa headed out the door. "I'll be back sometime tonight."

Rhona sighed. "Well, pooh." She sighed again. This wasn't going well. She strode to her bedroom and put on a brown jacket, turned, and stopped. She took off her jacket and threw it onto her bed. She sat next to it and immediately stood. She reached for her jacket then dropped her arms. With a loud grunt, she sat again.

Her mom stood at the doorway. "Oh, it's you. Everything okay?"

"Yeah, I'm fine."

"Of course you are. I know I always huff and grunt when I'm fine." She entered the room, sat next to Rhona, and rubbed her back. "What's wrong, honey?"

"Contessa's babysitting."

"Um, okay. Did you want to babysit?"

"No. I wanted her to go with me to ask if he would go back to the caves. If he says no, then I know he will say no, and once that equipment comes from Earth, it will be forever before we go."

"What? He who?"

"Fanar!"

"If Fanar says no then Fanar will say no?"

"No." Rhona groaned and rolled her eyes. "If Fanar says no, then Oswin will say no, and when they open the portal, we'll get all that equipment and Fanar will be extra busy and I won't be able to go exploring."

"Ah. Now you're making sense." Her mother rubbed Rhona's back in silence for several seconds. "Why do you think Contessa needs to go with you?"

"Mom, he'll listen to her because she's nicer than I am."

Her mom's rubbing stopped. "That's not true. You're just more expressive. You know what I think?" She kissed Rhona's cheek. "I think you're a bright and beautiful girl, and if you want to go exploring, then you'll find a way." She stood.

"Your dad should be getting home soon. Help me make dinner? Dr. Yamamoto has been working hard with some others on gathering food, and tonight we get to try it out. You've always been good at the culinary arts." Rhona spent the next couple of hours experimenting with the new food items as her mom mostly watched. She sampled each item and tried a variety of combinations until she was happy with the result. Satisfied with her finished product, she presented dinner to her mom and dad, who heaped praise on her ability to transform dull ingredients into an inspired meal. Beaming, Rhona went to her room and opened the drawer under her bed. Her next mission might require a bribe.

She strolled down her street to Main, followed the circle to E Street, and stopped outside the Monier house. His was the first on the street because it sat just below the meeting hall. The houses all appeared the same. Ninety round aluminum structures that had only taken two weeks to assemble. Rhona still marveled at how well they had been thought out that they could be put together so quickly. She took a deep breath, walked up to the house, and knocked.

Oswin's mom answered the door. In a hushed tone she greeted Rhona.

"Hello. It's Rhona, right? Come in."

"Hi, Dr. Carter." Rhona stepped inside. Dr. Monier slept on the couch. He had given up his bedroom when he took in the Carters. Oswin had once said Fanar's dad was so busy he would catch naps when he could.

"The boys are in their room."

"Thanks." Rhona took the long way around to their bedroom. It didn't feel right to pass the sleeping leader. She knocked lightly on the door.

Oswin opened the door. "Hi, Rhona. We were just talking about dinner. Did you try the new food yet?"

She followed him in while taking off her jacket and sat next to him on his bed. "We just did tonight."

Fanar grinned. "How'd Contessa handle it?"

"I don't know what she had. She's babysitting for the Stantons. Did you hear Mrs. Stanton's pregnant?"

"Wouldn't that make four children?" Oswin asked. "That's a lot of kids." "I liked the fruit," Fanar said. "What did they call it?"

"Roshberries. Little tart for my taste, but they're good," Oswin said.

Rhona reached into her jacket pocket. "So maybe you would like something sweet." She pulled out a chocolate bar. Three rectangular portions connected into a size that fit into the palm of her hand and were wrapped in paper.

Oswin laughed. "You weren't kidding about packing chocolate!"

"I would never joke about food." She opened the package and broke the bar into its twelve-gram pieces. After handing one to each boy, she put the last piece into her mouth. She closed her eyes and let it melt on her tongue. It had been a couple of weeks since she'd last had one. Contessa had guilted her out of three bars while she was sick, and her stock was dwindling, so she wanted to enjoy each bite to the fullest. She would definitely have to ask for more when she got the chance.

After the piece had melted to an almost indiscernible pad on the middle of her tongue, she swallowed and opened her eyes. Both boys were staring at her. "What?"

Fanar held up his hands. "Nothing. Just never knew chocolate could be such an experience."

Rhona gave a short laugh. "I bet you just horked it down and barely tasted it." *Boys. Gotta love 'em.*

Fanar blushed. "Well, I did enjoy it. Thanks." "I'm glad you did. I have a favor to ask."

"Oh ho! I knew something was up." Oswin laughed. "What do you want?"

"I want to go back to the caves. If we don't go now, we'll never go." She thought about saying more but didn't. She looked from Fanar to Oswin then back to Fanar.

"Well—" Fanar looked at Oswin.

"Please. This is my last chance before we get all those supplies from Earth."

"I'm supposed to be sequencing and cataloguing some new samples this week," Oswin said. "Maybe next week I can get two days."

Fanar replied, "I'm going to help Kellach analyze the electrical components of each building and catalogue anything we need from Earth. It's a lot of work."

Rhona's heart sank. She grabbed her jacket and stood. "Can you at least think about it?"

"Sure," Oswin said.

"Okay. Let me know as soon as you can." She had to get out of there. She forced a smile and quickly left the house. As she closed the house door, she blinked away the moisture in her eyes. *I am not going to cry. I am not going to cry.* She wiped away a tear. *Stupid tears.*

This is Fanar. In two weeks, the portal will open. Rhona wants to explore those caves before then, but I'm not sure I want to. Kellach's been spending a lot of time locked in a room of the maintenance building or with the Oracle. Helping him will at least help me learn more about this equipment. (System note: Day 47)

Kellach held a display in his thick hand. A cord tethered it to a circuit board inside the wall of the house. With the occupants of the houses mostly at work, it was easy for him and Fanar to do their diagnostics. Kellach disconnected the wire. "What unit is this?"

"C-7," Fanar replied.

"Right. C-7 needs one command node." Kellach handed the display to Fanar and put the wall panel back into place.

Fanar had been taking note of any additional items in need of repair. He nodded and entered the information in the hand-held pad. "Rhona's been asking if we can go exploring those caves again."

"I thought you were going to help me."

"That's what I told her. I said I have a lot of work to do." "Do you want to go?"

Fanar took a deep breath. "I don't know." Did he want to go? There were some things about those caves that puzzled him. The rock ramp up to the entrance, for one. There was also that sound he'd heard and the smell. Oswin was being coy about exploring with Rhona, not saying one way or the other. At first, Fanar had wanted to see if he could find an invisible creature, but now that he'd had some time away, he wasn't so sure about his idea. Of course, not finding his mom in the cave added to his reluctance. "I don't think so."

Kellach grunted. "That's probably the best. We need to check the solar panel output and batteries."

"Right." Fanar followed Kellach to the main power juncture of the house. A shadow passed along the wall. Fanar looked back to see one of the soldiers standing there.

"Kellach." The soldier waited until he had Kellach's attention. He glanced at Fanar. "I got it done."

"Good. That will, uh, be one less problem."

The man nodded and left.

Fanar watched the man go. The soldiers had nothing soldier-like to do, so they were used for a variety of tasks, and some of them went with others for tasks away from Landing. He'd thought they might just go back to Earth when the portal had opened a couple weeks ago, but they didn't have a choice. The government insisted on their presence. "What did he do?"

"Nothing you need to be concerned with." Kellach accepted the display from Fanar and connected the tether to a port on the battery pack. It showed the health of the batteries and other statistical information. "Batteries are still good."

"The Oracle ever come up with what made Contessa sick?"

Kellach stared at him. "That's the second time today you've asked." "That's right. Sorry." No, nothing new about what had made her sick. "I'm working with Dr. Huff on it. Why you so interested?"

"You're helping Dr. Huff?"

"Oh, well, you know. Making sure everything she needs is working properly. Me? I think finding this toxin is a top priority."

"Oh. Just curious, I guess."

"Still feel bad about that trip to the falls?" Kellach connected the display to the inverter. He tapped the screen, and it displayed the historical output of the solar panels. "Panels are good. See?" As he diagnosed the different items in each house, he made sure Fanar knew what was being done. Fanar already knew most of it because of his training on Earth. Kellach yawned then glanced at his uplink. "We've been at this eight hours straight. Let's call it a day and start again first thing tomorrow. We'll see if you've been paying attention."

Fanar nodded. "Sounds good." He packed the pad and diagnostic tools into their carrier and followed Kellach out the door.

As they neared the maintenance building, Kellach stopped and turned. "Here, let me take care of that. I'm headed that way anyway." He reached for the kit. "Tomorrow we can meet by the house we just left and go from there."

Fanar handed the diagnostic kit to him. "Okay." He watched as Kellach took the kit and continued on. He yawned and ambled to his house, the bright road panels lighting the area. Tomorrow would be a day of light. As he approached his house, he could make out Rhona's silhouette standing in front of it with her hands on her hips. He could only guess the expression on her face, but he was certain it wasn't good.

"Hi, Fanar," she said. Her greeting sounded pleasant enough. "How'd it go today?"

He stopped when he could see her face. She wore a neutral expression. "Okay. Spent the day with Kellach noting some things that need to be replaced and calibrated some other items so they work better."

"He's a bit odd."

Fanar smiled. "Yeah, but he's pretty smart. I'm actually learning a lot about the technology. What'd you do today?"

She frowned. "Contessa convinced me to help teach the younger kids with her and Miss Shaw." She dropped her hands.

A smile tugged at the corner of Fanar's mouth. Maybe she wasn't frustrated with him after all. "How'd that go?"

"I don't think that's for me, but that's not why I'm here."

*Darn.* "What's up?"

She put her hands back onto her hips with her fingers facing back. "You haven't given me an answer. Can we go exploring those caves again before they open the portal?"

She wanted an answer, and he wasn't ready to give her one. At least not one that she would like. What could he say? 'Sorry, Rhona, but I don't care about those caves and I just want to go home?' He could, but then the next two weeks would be miserable. "Um. Have you checked if there are any rovers available? I know they're used a lot more now."

Her face brightened and her hands dropped again. "No. Let me do that and I'll let you know. Thanks!" She ran off the road and toward the maintenance building.

Fanar went into his house. With any luck, there wouldn't be any vehicles available until after he was busy with the delivered equipment. He searched the cabinets, but there weren't any more prepackaged meals. That was okay; he would eat some of the locally harvested food. He grabbed the Roshberries and a small bowl. The berries grew in clusters like grapes but resembled strawberries without the external seeds. He set the bowl on the counter and pulled a berry off the peduncle and popped it into his mouth.

Oswin walked in from outside and smiled when he saw Fanar. "You sure made Rhona happy." He reached out and pulled a Roshberry from the cluster.

Fanar spit the pit into the bowl. "I did?" He put another berry into his mouth. Oswin squinted as he bit into the berry. "Yeah. You said we could go exploring those caves."

"No, I just asked if there were any rovers available. I didn't say I would go."

Oswin stared at Fanar for several seconds before speaking, his voice rising. "Sometimes I don't get you. Why would you give her false hope like that? What were you thinking?"

"I don't know!"

"Well, now she thinks that we're going to go exploring those caves again! What are we supposed to do if a rover's available?"

Fanar closed his eyes and put his hand on his forehead. He should have thought it through. Of course she would take that as affirmation to explore. "I don't know." He took his hand down. "I'll deal with it."

Oswin spit the pit of his Roshberry into the bowl on the counter. "You better make this right. Last thing we need is to be stuck here with some angry girls."

Fanar sighed. *How am I going to get out of this?*

# CHAPTER SIXTEEN

## Missing

**F**anar stared outside. In the distance, heavy clouds painted the sky an ominous grey. Besides the cirrus clouds that had been the mainstay of his time on Niton, he hadn't seen any clouds. During the preparations on Earth, the trainers had said it probably wouldn't rain at Landing, but doubts were brewing in his mind. He tore himself away from the window and went to the dining room. For breakfast, Oswin's father, Dr. Garrick Carter, had cooked some roots and covered them with a sauce prepared from another plant and water. Meant to resemble oatmeal from Earth, it didn't taste anything like it. Fanar took a bite and grimaced.

"What? Not to your liking?" Dr. Carter asked.

Fanar watched Oswin shovel a generous portion into his mouth and barked a laugh.

Oswin swallowed. "This is great!"

Oswin's mother, Dr. Candice Carter, pushed her half-eaten bowl away. "I have to agree with Fanar on this one. This will take some time to get used to." Fanar asked, "Have you seen my dad?"

She looked at the empty couch in the next room as if expecting to see him there. "I haven't seen James in a couple of days now."

Fanar frowned. While his dad would be gone for two or three days at a time, he would generally tell Fanar before he left for an extended time.

After eating, Fanar went to his room to see if he had any rain gear. He didn't. If it rained, the day would be miserable.

Oswin entered the bedroom and said as he rummaged through his clothes, "Now that is what I call service."

"What's that?" Fanar guessed Oswin was looking for rain gear. "One of the soldiers dropped off the rover we reserved."

*I never said I would go to the caves!* "What? I don't—"

"My parents and I are heading west along the lake today. There are some small organisms living in the shallows. We're going to trap a couple and study them."

"Oh." Fanar sighed in relief. "Guess those soldiers are really bored. Do you have any rain gear? I don't."

"No, I thought it wasn't supposed to rain here." Oswin pulled out a light sweater. "Here it is." He shoved some stuff back into the drawer and closed it. "See you tonight." He departed, leaving Fanar alone.

Fanar went outside. The sky overhead was bright blue. If it weren't for the ominous clouds coming in from the west, it could be easy to believe it wouldn't rain, but the air was even more humid on his skin and a cool wind caressed his face. Thankfully, he would be inside most of the day. He found Kellach waiting for him at the designated house.

"All set?" Kellach held the diagnostic kit out for Fanar.

As Fanar reached for the kit, he saw a grey curtain of water descending in the west.

"Huh. Guess they were wrong and we'll get a chance to use our rain collectors. Best we get moving," Kellach said and walked to the next house they were scheduled to examine.

Fanar watched the rain in fascination. Kellach leaned out the door. "You coming?"

"Yeah." Fanar followed him while he craned his neck to watch the rain.

The mass of water seemed alive, waving like fabric in a breeze. "Fanar."

"Sorry." He followed Kellach to the command console and opened the kit while Kellach removed the wall panel. "Think they might connect the buildings to the Oracle? I heard someone say we're supposed to be getting some fiber in the next few months."

"Don't know. I haven't been given any new information." Kellach held his hand out for the display.

Fanar handed the device to him. "Have you seen my dad recently?"

Kellach stopped. "Not recently. Why do you ask?" He connected the wire from the display to the command node's circuit board.

"I haven't seen him in a couple of days, and he didn't tell me he was leaving."

Kellach grunted. "Everything's okay here." He handed the display to Fanar. "Your dad's often gone for several days at a time."

Yes, Fanar knew that, but something felt different. His father's confession down at the lake had been the start of a more open relationship between Fanar and his dad.

Kellach measured the health of the battery pack and the solar panels and held the display out. "Fanar."

"Oh, sorry." Fanar took the display and put it into the kit. He walked to the next house and waited as Kellach knocked before entering and repeated the process. Fanar tried to remember the places on Niton his dad had told him about. Several sites had been set with small traps and cameras for wildlife observation. He could be at the mining site or with those gathering more food.

Fanar pictured his dad up at the site where the Roshberries were growing, directing people to not take too many so they would continue to self-seed and grow. He smiled to himself. That would have been his mom, really. His dad had more celestial thoughts, but he was the leader and very knowledgeable in many things. He very well could be doing that. After working with Kellach, Fanar would ask around. Surely someone knew where his dad had gone.

Fanar walked into the building that housed the Oracle. Electronics for the general populace were limited because of the scarcity of material; however, the government made sure the colonies had the best computer core.

Dr. Huff sat at a terminal, frowning at the display. "Oracle, show me all the sources we've tried."

The display blanked and then showed a scrolling list of chemical formulas and the named sources with pictures.

Fanar smiled at how she called it 'Oracle.' He recognized some of the sources on the display. Various plants that grew in and around the lake and some of the animals and insects in the region. "What are you looking for?"

Dr. Huff jerked in her seat. "Fanar! You startled me. I don't know. We've been working on this for over a month and haven't found anything to suggest what made Contessa sick. Maybe I'm being obsessive, but I like to know these things."

Fanar stared at the display. "So you don't know what the toxin is."

She waved her hand at the display. "Oh, Susan identified the toxin a couple weeks ago."

"Susan?"

"My nurse practitioner. She's been working with me for a few years. When I went back to Contessa's original blood sample, I found trace amounts of the toxin. Now we're just trying to identify the source so we can at least warn people. Ideally we can come up with an antitoxin."

"Trace amounts?" He scowled.

"Yeah, almost missed it. Your trip to Earth would have been fruitless. The sample I gave you and Oswin was after her body had eliminated it. If I had known she had such a small amount in her body, I would have sent the original blood sample, and then we wouldn't have found anything here on Niton." She resumed working on the console then turned back. "I'm supposed to be keeping your dad up to date. Do you know where he is? I haven't seen him in a couple of days."

"No. I hoped you would know."

"It might be good to ask Dr. Yamamoto. She knows the rotation for the food collection best. He could be with one of those groups."

"Okay, thanks." Fanar left the building. The dark clouds cast a gloom on Landing. The leading sheet of rain moved over the lake at a snail's pace. He shivered and rubbed his arms. It hadn't been this cold earlier.

He strode to the food storage building. It had originally housed the packed food containers brought from Earth. Teams gathered food on rotation from a variety of places. He opened the door and paused. Like a cloud, the rich aroma of earth encompassed him. It reminded him of when his mom would ask for his help in her garden. The dirt would get under his fingernails, and it would take a couple of days to get them clean.

Five people were assembled near the sink a few meters to his left. Three were taking leafy plants from collection containers and placing them onto the tray attached to the sink, while another sprayed them down before the last person moved them to a shelf to remove the excess water. Eating plum- colored leaves didn't appeal to him.

Near the sprayer, Dr. Yamamoto turned with a plant in hand. "Fanar. Here to lend a hand?" She put the plant onto the tray.

"No, I'm looking for my dad."

She held a finger up and went to the sink to rinse her hands off. She grabbed a cloth towel and dried her hands before walking over to Fanar. "Sorry, I couldn't hear you over the sprayer."

"Have you seen my dad recently?"

She glanced at the team as if confirming he wasn't among them. "No. Is he supposed to be with me?" She wiped some water from her forehead.

"I don't know. I haven't seen him in a couple of days."

Dr. Yamamoto scowled. "Well, I don't recall him saying anything to me about an extended trip, but I've been away. You sure he isn't around?" "I'm sure." He sighed and looked past her to the storage rooms. The

transparent doors allowed him to see inside. Most of them lay empty. He hadn't realized they were so tight on food. This whole colonization effort had become a disaster.

"Hey." She put her hand on his shoulder. "I'm sure he's fine. He'll probably show up any minute to get out of the rain." She looked toward the mountain as if she could see through the walls. "Funny, that. I didn't think it was supposed to rain at this location." Fanar nodded.

She dropped her hand. "While you're here, want to help?"

Helping would get the plants into storage faster so they could gather more food. "Maybe later. I gotta find my dad."

"Okay. Feel free to stop by anytime. We can always use the help." She went back and grabbed another handful of clean leaves from the tray and assembled them onto the shelves.

He went outside and heard his name called. He saw Rhona waving at him from down the road.

She ran up to him and smiled. "Perfect! I was just looking for you. I talked to Kellach and a couple of those soldiers—"

He held his hand up. "Not now, Rhona." "But—"

Fanar raised his voice. "I really don't have time!" He moved away from her. Who should he talk to next? "Don't you turn away from me!"

He shook his head and faced her. "I'm—" He stopped. Maybe their dads were together. "How long has your dad been away?"

Her eyes narrowed. "What?"

"When did you last see your dad?"

"This morning," she barked back. "He was going to stay at the mining site, but he came home. What is with you?"

*Maybe they came back together.* "Okay, thanks." He spun around and ran toward his house. He pumped his legs until he could hear his heart beating in his ears. With each ragged gasp, the cold air gripped his lungs with icy tendrils. He ignored the feeling. When he finally got to his residence, his feet were numb from the pounding.

Hopeful, he opened the door and stumbled inside. "Dad?" He coughed between labored breaths. The couch was empty. He glanced in the kitchen area, but it was empty. He coughed again. He went around toward the bedrooms calling for his dad. His dad may have given up his bedroom for the Carters, but there were still some items he would need from there on occasion. Empty. He coughed a few more times and laboriously took in some more breaths while he walked all the way around back to the living room. Besides him, the place was empty.

*Plink. Plink. Plink.* The aluminum structure received the first few raindrops. Soon, it sounded like he was back at the waterfall. The rain had arrived. His hope fell like the rain outside. What if his dad had gotten hurt? His eyes went wide. What if that invisible creature had gotten him? Taken him away like it almost took Contessa? He ran to his bedroom and opened the wall cabinet. Nothing. Nothing useful for rainy weather. How could they not pack anything for rain? He sprinted to the front door, thinking quickly. Who could he talk to? Kellach. Maybe he would know what to do. Fanar opened the door and hesitated.

The rain crashed against the bright road panels, the air fragrant with petrichor. He made a mad dash toward Kellach's house. Water ran from his hair into his eyes, making it difficult to see. He shook his face to clear his vision. The wetness plastered his clothes against his body. Wiping his eyes, he pounded on the door.

Kellach opened the door. "Fanar? What are you doing here? C'mon in and get out of the rain."

"Thanks. I think he—" Inside the house were several of the soldiers, including Randy. Fanar ignored them. "I think my dad's been taken by a native creature that's invisible." He knew without seeing the expression on Kellach's face what it sounded like. Even he wasn't so sure about it.

Kellach remained frozen for an agonizing two seconds then laughed.

When Fanar didn't laugh, he stopped. "You're serious?"

"You know how an octopus can change its color and skin texture to blend in?"

"Yeah." Kellach glanced at the soldiers. "Go on."

"Well, Oswin and I think that may be what's happening here. Remember how Contessa said someone with bad breath tried to take her? Well, I think it was this creature because I've also been smelling something, and when we were in the caves last time, I heard a sound like someone walking around, but I didn't see anything." He took a deep breath.

Kellach waited for a few seconds before responding. "That's it?"

"I know it sounds crazy. Honestly, I didn't know how crazy until I just said it out loud, but I just know there's something out there. Something that doesn't need to live in buildings and doesn't want us here. I think that this creature's been the one sabotaging our equipment, and when that didn't get rid of us, it was going to take Contessa, but we scared it away and now it took my dad. Please, Kellach, you gotta help me find him." Fanar's heart pounded in his chest as he waited. He glanced from face to face hoping to see something, anything, that would reassure him.

Kellach scratched the whiskers on his chin while staring at Fanar then looked outside. "I hope you don't want to go now."

Fanar sighed in relief. Maybe Kellach would help. "No—"

"And what's your game plan? Just wander around until we trip over this creature? The one that's invisible?"

He hadn't thought that far ahead. He stared at the floor and rubbed his forehead. Where would it be? He brought his head up. "By the lake. I've smelled it by the lake a few times. I think it lives there."

"The lake, huh? Okay, tell you what. If it's not raining tomorrow, we can dispense with the home diagnostics and see if we can find something." If they could find this thing, then they could find his dad. Fanar just knew it. And, perhaps, his mom too.

# CHAPTER SEVENTEEN

## The Hunt

*This is Fanar. After convincing Kellach to help, I've been recruiting others to find my dad. Actually, that's not true. I've been trying to convince people to help find an invisible creature that I think kidnapped my dad. I should have asked them to help find my dad; it would have been a lot easier. It's late, so we'll start first thing tomorrow. (System mark: Day 49)*

**F**anar lay in his bed listening to the rain. His dad was out there in this, maybe tied up near the lake. Eager to begin the search, he sat up. "Computer, lights at ninety. C'mon, Oswin. Let's go."

Oswin groaned. "Oswin, get up."

"Fanar, I got home late, and I just fell asleep."

"My dad's out there." He reached over and shook Oswin. "We gotta go."

"I'm up. I'm up." Oswin rolled toward Fanar and yawned. "So what's the plan?"

"Find the Nitonian so he can lead us to my dad." "Okay. I know that, but how? What's the plan?"

Fanar shifted on his bed. "You're the biologist. Isn't there a way we can detect chemicals in the air? We could look for a shift."

"You want us to go around sniffing the air? What if it lives in the lake?" Fanar's shoulders sagged. "I hadn't thought of that." He sat up straight.

"I don't care; we gotta try. My dad's out there! We gotta find him!" Oswin sat up. "I know, and we will. I'm just thinking about how." "We could always use the thermographic sensors in the rovers."

"True. Hopefully it isn't cold-blooded." Oswin stood. "Can we get some breakfast first?"

Fanar waved him toward the door. "You go. I can't eat."

"It's going to be a long day. You should get something in you."

Fanar shook his head. "Maybe later." His stomach felt like a lead ball. His eyes stung as tears formed. He wiped his eyes and stood. This would be the third day his dad had been missing. Or maybe he'd been out for over a day doing something and only recently the Nitonian had taken him. If that were the case, his dad might not be in such bad shape. He closed his eyes. *Oh, please let that be the case.* Fanar was just getting his dad back; he didn't like the idea of him being gone. He opened his eyes and stared out into the dark. The light from inside the house reflected off the rain. This wasn't going to be easy.

Fanar found Oswin and his bleary-eyed parents eating breakfast. They were eating more of the cooked root and sauce. There was no way he could choke that down. "Thanks for helping me find my dad."

Oswin's dad waved it off. "He's a good friend. We would do anything for him. It is unusual for him to be out like this, but I'm sure he's fine. He didn't say anything at all about being away?" Seeing Fanar shake his head to the negative, he continued, "Nothing to us either. We'll probably find him in a rover at one of the research sites sleeping off a long day of work. However, it's still best to find him."

"I'll go get a rover while you eat."

Oswin's mom stood. "Hold on; you can take a raincoat." She went back toward her bedroom.

Oswin shoveled in another bite and talked as he chewed. "They had some in the rover yesterday. Mom put them into her bathroom to dry when we got home. All this rain. Pretty amazing, huh?"

Fanar accepted the coat when she came back. "Thanks." He opened the door and went out. The rain was as intense as the previous day, with no end in sight. His breath fogged and hung around him as he walked to the maintenance building. He didn't even check the time before leaving. He frowned at the empty place on his wrist. Oswin had said he had just gotten to bed, so it must be very early, which would explain why the streets were deserted. Hopefully the others would be joining him soon.

He continued around to the back of the maintenance building and tugged at the entrance by the bay door. It was locked. When had a lock been installed? He wound his way back to the front entrance and found it locked too. Clenching his teeth, he balled his hands into fists and shook them. *Why are these locked? Kellach should have the courtesy to give me access! Am I not his assistant?*

With a huff, he marched to Kellach's residence. Kellach knew about the search; why would he lock the doors? When he got there, he pounded on Kellach's door.

Eventually a soldier opened the door, yawning. "Fanar, right?" He waited for acknowledgment, which came as a head nod. "I think Kellach went to get the rover. He's probably headed to your place now."

"Thanks." Fanar jogged to his house. How many soldiers were there? Five or six—he couldn't remember. When he got to his house, a rover was sitting there. He went inside the house.

"There you are. Ready?" Oswin asked and went outside to get into the waiting vehicle.

"Let's go find your dad," Oswin's mom said. Oswin's dad followed her, leaving Fanar standing in the house.

After going outside, Fanar got into the front seat and found Randy there. "Where's Kellach?"

"He had some other business he had to attend to. So where to, chief?"

*My dad's not important?* Fighting down the urge to scream, Fanar gave some direction. "Let's start on the east side of the lake and move westward."

"You got it. Kellach said you wanted to use the thermals?" "Yeah."

They drove to the east side of the lake. Rain pounded the top and sheeted off the windshield. When they got close, Randy turned west and activated the thermographic sensors along with the full set of lights.

Fanar told Randy, "You probably think this is a waste of time."

Randy shook his head. "Not for me to say. I never knew an octopus could camouflage itself. After you left, my brother and I went to the Oracle and inquired about it. Saw a video that rocked my world. We have some incredible technology in the army, but nothing we have comes close to that." *He must have been really curious to go through the rain to watch a video.* Fanar watched the screen. The dark blue area on the left was the lake. Next to it, the ground showed as mostly light blue with blotches of yellow moving to the right. He looked up to see small creatures scurrying away from the moving vehicle in the rain.

A hand patted his shoulder. "Don't worry, dear. We'll find your dad," Oswin's mom said.

*How?* If they found this creature, how would it lead them to his dad? What if it lived in the water like Oswin had suggested? Would it drag his dad down there too? A dark orange dot appeared on the display and got larger until it dominated the screen. He looked up and saw a boulder. Randy drove around it.

"Radioactive material," Oswin's dad said. "We'll have to remember where it is so we can avoid it. Maybe after we get settled in, we can use it in a reactor, but for now it must stay put. Strange though, I don't recall any reports of radioactivity from the drones."

Randy nodded and hit a button on the screen to save the location.

Fanar stared out into the black world and sighed. Convincing people there was an intelligent creature that could blend into the environment was a harder sell than he'd thought it would be. The drones had taken hundreds of hours of video and thousands of snapshots in a wide area when the planet was originally researched. Nothing had indicated a civilization or intelligent life, so naturally they didn't accept his theory.

"How far you want me to go, chief?"

Fanar swung his head forward. "All the way to the mountain on the far side of the lake."

Randy nodded and shifted in his seat. It was going to be a long day.

When they got to the mountain, the rain had stopped and Oswin's dad suggested a break. "We've been at this for several hours, and I need to stretch, amongst other things."

As they got out of the rover, another one pulled up alongside them. Another soldier and Dr. Yamamoto got out, along with Contessa and Rhona. After Rhona got out, she turned on Fanar. The lights reflecting off the mountain illuminated her determined face. Her hands went on her hips. "How could you do that?"

Fanar stared at her, confused. "Do what?"

"You didn't tell us you're searching for an invisible creature. How could you just leave us behind like that? You're lucky we ran into Dr. Yamamoto so we could help." "Look—"

"No, you look! From the day we met, you've been condescending and rude to us, and I'm tired of it. You're just lucky that we're trying to find your dad, or I wouldn't be helping at all!"

Fanar knew that wasn't exactly true. The idea of discovering something new would be enough to make her come, but there it was. His gaze went from person to person. They probably didn't believe in the creature; they were just helping to find his dad. He didn't care as long as his dad was found.

Contessa shook her head at Fanar and went to the other side of their rover away from him. Rhona dropped her hands and followed her sister.

Oswin came up next to Fanar. "You didn't tell them what we're doing?" "I—" Fanar shook his head. "I just didn't think about it! Okay?" He stormed away from Oswin and went to the edge of the lake. He was on the right path; he knew it. He put his hand to his forehead. Ripples in the water reflecting moonlight grabbed his attention. Turning around, he saw the clouds had parted to show a half-moon. He wasn't sure which one. When he lowered his eyes, he gasped. Contessa was standing two meters in front of him.

Contessa approached him. "She gets rather excited. Sometimes you have to just let it slide off your back." She looked at the water. "Isn't there a river that pours into the lake this way?"

"Yeah, two actually. One almost two kilometers down the face of the mountain and another one farther down."

She nodded. "You really think it's out here?"

"I do. It's the only thing that makes sense. The damaged equipment and everything that's going wrong. We're the invaders, and it wants us gone."

"That's scary. The thought that something so dangerous can be next to you without your knowledge. What will you do if we catch it?"

*I can't come up with all the answers!* "I don't know. Somehow, we gotta find a way to communicate with it. Find out where it took my dad." *And hopefully find him alive.*

"Don't worry. I'm sure we'll find him. I better get back. Rhona wants to be the first to find the Nitonian."

Fanar watched her walk to the rover she'd come in. After a minute, her rover moved north, parallel to the mountain. He strode over to Oswin. "You ready?"

"Yeah, let's find your dad."

They climbed into the vehicle and drove east in a path parallel to the one they'd originally taken. After they traveled almost twenty kilometers, they passed a boulder that resembled the one they had seen earlier.

Fanar scowled. "I thought we were farther from the lake and wouldn't see this again."

Randy tapped the screen. "No, that's a different one. We're about five klicks north from where we saw the first one."

It seemed exactly the same to Fanar. He stared at it as they passed. The pale moonlight made it difficult, but he continued to look at it while they drove. Wait, did that thing move? "Stop!"

"What?" The vehicle stopped. "Turn around! Go back!"

"What's going on, Fanar?" Oswin's dad asked.

"I think I saw it move. Go slowly."

"You got it, chief." Randy circled the vehicle around and inched it forward until they came to the boulder.

Fanar leaned forward. The boulder was there, unmoving. He opened the door.

Oswin's dad put his hand on Fanar's shoulder. "Fanar, that thing's not safe. See the heat it's generating?"

"I don't think it's a rock." He stepped out of the rover and approached it. The others came with him and surrounded it. Nothing. It didn't move a centimeter. Oswin looked at him and shrugged. Fanar held up his finger and stared at the boulder.

He hadn't imagined it, this thing had moved. Should he touch it? He was picking his foot up to step forward when the top of the boulder collapsed. He froze. Randy raised his ever-present rifle. Fanar held up his hand. The last thing he needed was to have it shot. The soldier was probably even more trigger-happy than normal from lack of activity.

In another ten seconds, the sides had sunk in half a meter, and the top came down to half its height. Fanar took a couple steps back, his heart pounding as he watched this dramatic change.

Simultaneously, six tortoise-like legs became visible as the body of the creature rose and a melon-sized head atop a worm-like neck appeared with eyes facing forward. The head ascended, and soon two arms, attached to the base of the neck, were visible with four tentacles at the end of each arm. The tentacles were easily long enough to wrap around a person. When it had finished its transformation, it stood nearly two meters tall. The bulk of the body was about a meter high, less than a meter wide, and a little longer than a meter front to back. The surface of the body resembled a wrinkled prune. It spun around and stared at Fanar.

Fanar stumbled back when he saw the eyes. They were large red eyes with yellow thorn-like protrusions in a spiral pattern around a dark yellow iris. An urgency to run nearly overwhelmed him.

The creature made a whistling sound, some guttural noises, and some clicks through its mouth as it turned all the way around. Before anyone could react, it closed its eyes. After a few seconds, sparks leaped across its body. Fanar jumped back, unable to tear his eyes away. It got brighter, and soon he couldn't look anymore. He closed his eyes, and in a second, the intense light went away. He opened his eyes to see the others staring back at him. The creature had disappeared.

"Oh! What is that smell?" Randy backed up, waving a hand before his scrunched face.

Fanar knew the odor. It was the same as in the caves and by the lake with his dad. It was the odor of an invisible Nitonian.

# CHAPTER EIGHTEEN

## Capture

**F**anar rushed into the empty place. It was truly gone. He hadn't heard it go away; the soft ground must have masked its retreat. His heart pounding, he looked around at the others. "What was that?!"

Oswin's mom had her hand to her chest, which rose and fell quickly as her mouth moved silently.

After several seconds of voices overlapping each other in excitement, Randy's voice rose over the others. "Quick, get into the rover!" He ran headlong to the vehicle. "Maybe we can track its movements with the thermal sensors."

Fanar scrambled with the others to the rover. He stared at the display, watching for any indication of the creature. If they could track it, they could find it and maybe find his father. After a second of watching and seeing only the light blue of the ground and occasional yellow dots of smaller creatures, he came to the only conclusion he could. The creature was not there. He had seen it. It was real. Loud breaths distracted him from his thoughts. It took a second before he realized it was his own breathing. He took a deep breath followed by another to regain control. This thing not only camouflaged itself, it could make itself actually invisible!

Randy moved the rover into a widening circle. After several minutes of searching, they could not find the creature. He stopped the vehicle. "Okay chief, this is what I think. This thing is either invisible to our heat sensors too or it can move quickly. I say we go back to Landing and tomorrow get all six rovers engaged to find it instead of just the two."

The plan sounded good, so Fanar agreed. Maybe now others wouldn't be so hesitant to join. He wasn't sure what he'd expected, but this was beyond his wildest dreams.

"I can't believe what we saw!" Oswin said.

"When that head came out, I just about fainted," Oswin's mom admitted. While they returned to Landing, Fanar kept his eyes on the display, hoping to see the creature.

This is Fanar. I can't sleep. I don't think I've slept for two days. My dad's been missing. Did I say that previously? I don't know. This Nitonian took him. When everyone's available, we will find this creature that can become invisible and somehow get it to show us where it took my dad. I can't lose him too. (System mark: Day 50)

Oswin's face stared into Fanar's. "Fanar, you awake?"

Fanar found himself in his room, on his bed. He didn't remember coming back. "I'm at the house."

"Yeah. I woke you up two minutes ago, but you were incoherent." Fanar forced air into his lungs and released it.

Oswin waved his hand in front of Fanar's face. "You with me?"

"Yeah. I'm"—Fanar slid his hand down his face—"I'm here." He took another deep breath.

"So what's the plan today?" Oswin sat on his bed.

Fanar rolled onto his side. All night, he had been thinking about what they should do, his mind running around in circles, chasing fleeting ideas. After they'd returned to Landing, he'd wandered around, playing the last few days over and over in his mind. He remembered making a log entry on the Oracle but still wasn't sure how he'd gotten in his bed. He couldn't remember what he'd said in his log entry either. "I've been thinking, now that we've seen it, a lot more people might be willing to get involved in finding it."

"So you think it's still around?"

Fanar numbly nodded. His whole body dragged from mental and physical exhaustion. "I'm just not sure how to capture it. Maybe we can use ropes. We can check storage for some."

Oswin stood. "Sounds good." He grabbed some clothes from his bed drawer. "Let's see who's here."

"Here?"

"Yeah, my mom said we had visitors five minutes ago."

After getting dressed, Fanar and Oswin entered the living room. Fanar slowed as they passed the area where his dad slept. Whatever happened, they'd better find him. Waiting for them were Contessa and Rhona.

"I can't believe you found it first!" Rhona exclaimed. She was sitting next to Contessa. She leaned forward. "What was it like?"

"Well—" Fanar started.

"I wasn't talking to you," Rhona snapped. He flinched at her outburst.

She narrowed her eyes at him for a second before focusing on Oswin. "So?"

Oswin glanced at Fanar then responded to Rhona, "Like a giant slug with six legs and arms."

Rhona grimaced. "Ewww, gross."

"I thought of it as more like a six-legged tortoise with arms and tentacles," Fanar said.

"So it can become invisible?" Contessa asked.

Oswin glanced at Fanar. "I talked to Dr. Huff about it. It got bright and we couldn't see it. She thinks that maybe it has some sort of intense bioluminescence that made the flash and a natural camouflage that would allow it to get away. It was rather dark." He looked at Fanar again.

"I just hope we can catch it today," Fanar said. Invisible or just camouflaged, he wanted to find out what the creature had done with his dad. "Let's see if we can find Kellach. Maybe he has something that would help."

Contessa stood and put her hand on his arm. "We're here to help in any way we can."

As Fanar walked ahead of the others, he moved his shoulders as if he were shrugging off a coat. He could almost feel Rhona's icy stare from behind. He hadn't meant to upset her by searching for the Nitonian without telling her. He knew she wanted to discover something new here, and now she was upset because he'd found it and she hadn't been there. He sighed and did his best to dismiss it. Once they found this thing, Rhona would be better.

Oswin ran up to Fanar's side. "So the way I see it, this Nitonian likes to stay near the lake. Do you know if we have any hand-held thermal cameras?"

Fanar looked at Oswin's face in the glow of the street panels. The sun would be rising later in the day. "I don't know. I can always ask."

Contessa asked Rhona, "Doesn't Daddy have some sort of special camera?"

"How am I supposed to know?" Rhona asked.

"I just thought you might know." Contessa frowned.

At the meeting hall, over twenty people stood outside waiting for them. All six rovers were nearby ready to join the search. As Fanar approached, several people rushed toward him to get their questions answered.

"How big's the creature?"

"I heard it can turn invisible." "You think it has your dad?" "We need to study it."

Fanar hated being the center of attention. Like the creature they'd found the previous day, he wanted to hide and run. He studied those gathered. "I don't know how, but we need to capture it. I really believe he has my dad." One of the soldiers shouldered his way through the crowd, carrying something that resembled a large double-barrel rifle with a net folded between the wide-spread barrels. Each barrel tip had a weight attached to the net. His face was set with a grim determination while his eyes gleamed.

Probably excited about having something to do. "We have several of these net guns. If we break into six teams, each can have one."

The voice of Terrell King, Contessa and Rhona's father, came from behind Oswin. "Hopefully the net can hold it."

"Daddy"—Contessa turned toward him—"don't you have a special camera that can help us find it?"

"I have a spectroscope for analyzing minerals. You're probably thinking of a thermal imager. I don't have one of those."

Her face fell. "We're not going to find it, are we?"

"We'll give it our best shot. Can't ask for more than that."

The crowd broke into small groups of four or five and moved toward the rovers. Each rover had a soldier driving it.

"You ready, chief?" Randy approached them with a net gun in hand. Oswin craned his neck around. "Have you seen my parents?"

Fanar faced the direction of their house. "I don't see them. Where did they go?"

"I don't know. My dad said he may have something that would help with the search."

Kellach broke from the dispersing crowd and approached them. "That's some luck you guys had finding the creature. I think everyone else was hoping to find it first."

"Everyone else?" Fanar asked.

"Sure, many people were searching outside of Landing on foot and all the rovers were out. You left so early you missed it."

"I had no idea." Fanar took a shaky breath. He'd thought he was the only one who missed his dad.

"Hope that luck stays with you today." "Me too. Aren't you coming?"

Kellach pointed toward the back of the Oracle building. "I need to repair those water bladders. Two of them burst last night."

Oswin's eyes narrowed. "You mean the creature damaged them." "It looks that way," Kellach said.

Fanar scowled. "Was the portal generator damaged?"

"No damage seen." Kellach pulled his lips into a slight smile. "Besides, Earth will be opening the portal in a week, right?"

"That's my understanding, but not anymore after that."

"What?" Rhona asked. "I thought we were opening the portal. And what if something happens here and we can't open a portal back?"

Oswin grinned. "Then we're stuck!"

Her eyes went wide. "Seriously?"

Fanar sighed and shook his head. He was not in a teasing mood. "No. If a portal isn't opened from this side at the scheduled time, then Earth will open one within a month."

"Awww, you should've let her sweat it for a bit," Oswin said. He saw his dad. "Hey. Did you find what you were looking for?"

Oswin's dad held up a slim box with antennas of different lengths sticking up from the top. "Sure did. I thought maybe I left it on Earth because it wasn't where I remember putting it."

"Where's Mom?"

"She's staying away today." He lowered his voice. "I think yesterday's adventure scared her."

Fanar stared at the device Oswin's dad held for several seconds. "Is that a directional EM analyzer?"

Oswin snorted. "Of course you would know what it is."

Oswin's dad nodded. "Yeah, not sure if it will help or not, but thought we could at least try."

Fanar went to the remaining rover. "I just hope it isn't too late." He climbed into the front seat.

Randy walked around and got behind the steering wheel. Oswin's dad sat next to Fanar.

Oswin, Rhona and Contessa got into the back seat. Contessa put her hand on Fanar's shoulder. "We'll find your dad."

Randy moved the vehicle forward. "I assume we're looking in the same area as the other day, so we've coordinated a search pattern."

Fanar nodded. "Okay."

"With all the rovers involved, we'll catch this thing," Randy said and switched on the thermal sensors.

They drove down toward the lake and turned westward like before. Fanar didn't look at the console; instead, he watched the horizon for boulders. Now that he knew about the Nitonian's ability to transform itself, it seemed the best thing to do. Far to the north, he saw the dot of another rover moving in the same direction.

"It would be easier if we could use radios to coordinate," Randy mused as he reached the mountain.

"Where do we go now?" Oswin asked.

"We go north and move back above Landing. The other rovers are doing the same." Randy redirected the rover northward.

Fanar stared ahead without saying anything. His mind went to yesterday and how the creature had stared directly at him. Like it knew him. He could swear there had been fear in its eyes.

"It's just a matter of time, Fanar. This search pattern is sure to flush it out," Oswin said.

Fanar tensed and sat upright. It hadn't just been fear, it had been panic! The creature had been in a panic. He turned around, got on his knees, and leaned onto the back of his seat to face Oswin. "It panicked."

"We could see that," Oswin said.

Fanar shook his head. "You don't understand; I could feel the panic. Somehow, I could feel it. I think we're going the wrong way. We found it near the lake last time, so we should go someplace where it would feel safe." He faced forward and sat. "We need to go down by the falls. With all the trees and rocks, it would be easier to hide."

"All that way? That's a long way to go. It would have to travel pretty quickly," Randy said.

Fanar nodded. "I can't explain it. I just know."

"If we're changing our search parameters, I need to let someone know." Randy maneuvered the vehicle so it was headed directly toward Landing. "Let me tell Kellach before we go to the falls so he can relay it to everyone else."

Fanar's eyes ached as he scanned the landscape just past the waterfall. His eyelids felt like they were being dragged over sand when he blinked. He was so tired that his body felt toxic. While others had dozed on the trip, he had stayed awake. He closed his eyes and rubbed the outer corners in an attempt to stimulate the tear glands. He kept thinking about how his father had spoken to him at the lakeshore. Now his dad could be gone.

He opened his eyes. Oswin's dad had directed the analyzer to the front of the moving vehicle. It showed the strength of various bands of electromagnetic energy. In the lower bands, the readout was practically nonexistent. These were the frequencies of the radio transmissions that were somehow absorbed, keeping them from using radio technology. Not that it mattered at the moment. Fanar would have been surprised to see something at those frequencies. Nothing stood out. Minutes later, a few lines jumped in the upper bands.

Fanar sat up. "Did you see that?"

Oswin's father nodded. "Yeah. Stop the rover." He moved the device to the right and then to the left. Lines leaped across the display. He held it still and watched the lines bounce.

"There's nothing on the thermals," Randy said.

Fanar looked at the thermal output on the console. True to what Randy had said, the display showed nothing outside the normal colors for plants and small animals.

"Stay here," Randy said. "I'll try to launch the net over it." "Over what? How can you tell where it is?" Oswin asked.

Randy gazed in the direction the analyzer was pointed. "How far away would you say it is?" Oswin's father gave a shrug as an answer. "Okay, we do it the hard way." He exited the vehicle then drew a line in the air with his finger to some distant point only he could discern. He walked to the tail end of the vehicle to retrieve his net launcher and worked his way back to the front, where he drew the line in the air again. Then he stepped forward three large strides and launched the net.

When launched, the net unfolded as it fell onto its target. In this case, that target was the ground. Randy pulled the line attached to the net and repacked the net onto the gun. He stepped forward five large strides and launched. Again, the net fell to the ground unhindered.

Oswin's dad scrutinized the output on his display. "It's a biologic. It can't be that far away. See how much energy is registering?"

Randy launched the net.

Perhaps the readings on the analyzer were from something else. Fanar was opening his mouth to say something when the net stopped midair and the sides draped down.

Fanar leaped across the seat to the door. "He got it!"

What happened next made his heart stop. Randy pulled the trigger a second time, sending an electric shock through the net. Fanar could practically feel his own skin burning. The Nitonian became visible and collapsed. The legs disappeared into the wrinkled folds of its outer skin as its head and arm protrusions thudded to the ground.

"Stop!" Fanar yelled and jumped out of the rover. "Stop! You're killing him!"

Randy pressed a button to stop the current and turned to Fanar. "You always incapacitate the enemy during capture."

Fanar ran past Randy. "We don't want him dead!" He stopped short a meter away. Was the creature alive? How could he tell? There, a movement. The creature moved his head. The eyes opened and looked at Fanar. No, they pleaded with him then closed again. *He's alive.*

"Is it alive?" Oswin came up next to him. Fanar nodded. "Yeah. He's alive."

Oswin studied Fanar. "He? And how do we know it's the same one?"

There could be tens or hundreds." He scanned the area as if others would appear next to them.

Fanar couldn't explain it. Somehow, he knew this creature lying in front of him was the same one that had been near Landing and in the cave too. Was it the same one that had tried to abduct Contessa? Maybe. "We need to load him into the back of the rover." He stepped up, peeled the net off and tossed it aside, watching for movement.

"Are you sure that's wise?" Contessa asked. She continued to sit in the rover after Randy moved the vehicle closer to the creature.

Rhona went with her dad past Oswin and stopped next to the side of the Nitonian. She studied it for several seconds as Fanar worked. "It has a spine like we do." She pointed to the neck. "See? You can see the bones under the skin."

Randy stepped next to her and handed out protective gloves to the men. "Let's see if we can lift it." He waited until it was surrounded. "On three, we lift." He squatted and slid his hands under the skin. "Is that a ribcage?" Fanar reached underneath. The skin was thicker and tougher than he had thought it would be. His hands met a firm structure. It could be a ribcage.

"I don't know. I think so."

"Could be; don't know without a proper exam," Oswin's dad said. "Okay, here we go." Randy started, "One—"

"Wait!" Contessa got out of the rover. "Someone should hold its head." She got near the head and stooped to grab it then stopped. She squinted her eyes and turned her face away as she put her hands underneath the watermelon-sized head.

"You okay there?" Fanar asked.

She kept her head turned away. "Yeah. I'm good. It just stinks."

Randy snorted. "Okay, on three." He counted up, and when he reached three, they lifted.

Fanar couldn't tell exactly how much it weighed, but it felt heavy. Probably close to a hundred and fifty kilos. They walked in step with each other and hefted the creature onto the floor of the rover's back segment with its head going in last. Contessa gently laid it down.

"We should put the net back on and secure it," Randy said as he retrieved the net.

Fanar opened his mouth to retort and stopped. Randy was probably right.

They didn't really know how dangerous this creature was, and if it had tried to abduct Contessa and taken his dad, they needed to keep it contained. He watched as Randy put the net on and locked the corners of the net to metal loops on the inside walls. The question now was, how were they going to find out where it had taken his dad?

# CHAPTER NINETEEN

## Interrogation

**"W**ell, how 'bout that." Kellach stood outside the meeting hall looking at the Nitonian. The creature watched him. "So this is what's been causing all the problems around here, Fanar?"

"Yeah. He's also the one that took my dad. I know it." The creature moved its head and stared at Fanar. The area around the eyes relaxed, and it made a whistling sound followed by some clicks. It moved its tentacles through the net and reached toward Fanar.

"Watch out!" Randy shoved Fanar back and reached for the trigger of the net gun.

"Don't—" It was too late. Randy pulled the trigger, sending a burst of electricity through the creature. It shrieked and spasmed against the net before collapsing. Fanar unclenched his fists and opened his eyes. "You didn't have to do that!" The words tore out of him.

"He's obviously an aggressive creature. Twenty homes can't be lived in because of the destruction he caused. Water bladders ruined and lab equipment damaged." Randy's brown eyes stared into Fanar's own. "And didn't it try to abduct Contessa and then take your dad? Honestly, I don't know why we don't just kill it."

"If we kill him, we can't find my dad."

Randy glanced at the unconscious creature. "Well, either way, we had to subdue it before moving it."

Kellach yawned. Behind him, the stars grew dim as the sky brightened. It would be daylight soon. He ran his hand through his thinning hair. "Me? I think inside here would be ideal. There's a room that can hold him. It has no exterior windows, and it can be easily guarded."

With the decision made, the net was pulled away, and they moved it out of the rover and into the meeting hall. Contessa noted, "It doesn't stink anymore."

"Becoming invisible causes the odor. When he's visible, the smell must dissipate," Fanar said.

Kellach moved chairs and tables out of the way as they proceeded through.

"We should give it a name," Rhona said.

Oswin grunted as he carried the creature. "I was thinking about that. What do you think about Octoslug? Octo for short." He grinned. "Octo?"

Contessa studied the head she held. "Octo? I guess it could look like an Octo."

Fanar just wanted to sleep. If he could have gotten away with it, he would have lain on the floor. It took all his remaining energy to finish transporting the creature to the conference room. Once they had him locked up, Fanar was going to bed. He nodded his assent. For now, he didn't care.

Fanar stretched. He threw off the cover and sat upright at the edge of his bed. The ceiling's glow revealed an empty bed next to him. He showered and sauntered to the kitchen. A plate of food sat on the counter.

"We let you sleep," Oswin said from the living room. "You obviously needed it."

"Thanks. What's this?" He brought the plate up to his face and sniffed. It smelled sweet. One of the local roots and something else he didn't recognize.

"While we've been busy testing local lake organisms and minerals and such, Rhona's been experimenting with food. It's actually pretty good."

Fanar put a morsel onto his fork. "You also liked that mush your dad made." He nodded after taking a bite. "It's not bad." He took another bite and talked as he chewed. "Where is everyone?"

"Where do you think? All of Landing is stuffed into the meeting hall to get a glimpse of Octo."

"Octo? Oh, right. Octo." The name fit but didn't seem right. No, somehow the appropriate name would come about in due time. He let it go, for now.

"Several people are wondering how it escaped all the drone surveillance."

Fanar shoveled in another bite. Who knew Rhona was such a good cook? "I bet if we reviewed that footage, we'd see several boulders. Maybe that's how he sleeps."

"Then it must spend a lot of time sleeping, because we've never seen it before."

When they got to the meeting hall, Fanar could tell Oswin wasn't kidding. They hadn't encountered a single person on the road and nearly everyone inside was amassed near the front of the meeting room in an attempt to worm their way to the window of the conference room. Those who had seen it were huddled in groups talking about it or trying to get another glimpse. Some were curious, while others were angry.

Dr. Huff came alongside them. "The way you talked, I thought it would be bigger."

"You should see it when it's inflated like... like a pufferfish. It's almost twice as big," Oswin said.

"We've already collected some tissue samples. Everyone will want some time to study it, though I don't know how we'll get an x-ray."

"We have to find my dad first," Fanar said. "How do we get everyone out of here so we can try to communicate with him?"

"I'll take care of it." Dr. Huff worked her way toward the front and activated the audio system. "If I may have everyone's attention?" She waited for the room to quiet. "This is truly an extraordinary find, and I promise we'll have time to learn more about this creature. However, right now we have more pressing matters. Like the rest of us, Fanar is worried about his father. I would like the security team to remain, along with Fanar and myself. If everyone else could please go about your normal duties. Thank you."

Reluctantly, people filed out. Some patted Fanar on the shoulder and gave words of encouragement before departing. Soon only the handful of soldiers stood in the vast space with Dr. Huff, Fanar, and his friends.

Rhona glared at Fanar. "If you think I'm leaving after all we did to get it here—"

Fanar turned to Dr. Huff. "They should stay."

Dr. Huff nodded. "Okay, this is your show for now. How do you want to proceed?"

*How to proceed?* A big question he really didn't have an answer to. *How do you communicate with a creature from another world that doesn't know your language?* An image of his father flashed through his mind. He should start with that. "We need to show him a picture of my dad. Maybe some images of Landing and the area."

"Let me take care of that," Randy said. "I'll go to the Oracle and get a set of images loaded onto a data disc." He ran out the door.

Fanar approached the conference room and peered through the window. The others crowded beside him. The room was seven by five meters. Originally intended for meetings, its inaugural use was the occupation of a non-human. The Nitonian stopped pacing and looked directly at him while opening his tentacles and glancing around the room then back. Fanar got the sense he was asking about his confinement.

Kellach returned with Randy and whispered into the ear of one of the soldiers, who nodded. Kellach looked at the Nitonian and strode out.

Randy opened the door of the room as the security team readied their weapons and everyone else squeezed around the window. A large portion of one wall was a vidscreen with controls next to it. He put the data disk in and brought up an image of Fanar's father.

The creature glanced at Randy before returning his gaze to Fanar. The large colorful eyes gave Fanar a shiver. *Why is he singling me out?*

"Hey," Randy called out to the creature and pointed at the image. The creature remained still.

"He's only interested in Fanar," Oswin said.

"Maybe he knows that's his dad," Rhona suggested.

Randy tried again. He whistled at the creature. The creature's eyes went wide, and he shifted his gaze to Randy. "Yeah, you understand that, don't ya?" He whistled again and pointed at the screen. This time the response was a movement of the head back toward Fanar.

Fanar stepped toward the door. Obviously this Nitonian had it in for him. If anyone was to communicate with this thing, he had to be the one doing it.

"What are you doing?" Dr. Huff cried out.

Fanar turned. Contessa was biting her bottom lip. Rhona had put her hand on Oswin's shoulder. He locked eyes with Oswin. "It's gotta be me." After getting a nod from Oswin, he went into the room.

Perhaps the creature recognized Randy as the one who had shocked him. Fanar wasn't sure. But when he stepped into the room, the creature took a step toward him and stopped. The creature looked at Randy and then back at Fanar. The Nitonian's eyes followed Fanar's every movement. Fanar considered the bulk of the creature with its thick wrinkled skin and long tentacles. He could probably crush Fanar by charging and slamming him up against the wall or choking the life out of him. Fanar forced his legs to move him to the image of his dad and pointed. "Randy, do we have an image of Landing and the surrounding area?"

"Yeah, hold on." Randy changed the image shown to an aerial view. It was a stitched image of all the ones taken by the drones, encompassing Landing and the surrounding area. An area over one and a half million square kilometers.

This got a reaction. Fanar jerked as the creature raised his head a full half meter. The movement was nearly instantaneous. Much faster than Fanar would have guessed possible. He swallowed. The creature was actively studying the image. Did he recognize the area? "Show a picture of my dad in the corner." When the image showed, Fanar pointed at it. "My dad"—he moved his arm in a circular motion around the rest of the image— "where is he?"

The Nitonian stepped toward the image. Randy raised his rifle, and the other soldiers advanced. The creature stopped. Fanar believed that, if he tried, the Nitonian could take out all the soldiers with minimal damage to himself. Clearly, he could move very quickly.

"Maybe you should move aside," Randy suggested.

Fanar moved toward the window. Once away, the Nitonian took another step toward the image displayed. He peered at Fanar's dad then at the image of the area. He whistled and looked back at Fanar then back at the image before raising a tentacle and touching a spot on the screen. The tip of his tentacle was on Landing. Fanar scowled. He knew his dad wasn't at Landing. Nobody had seen him for days.

"Maybe he misunderstood," Contessa said.

Fanar approached the image, ignoring Randy's warning, and pointed to the photo. "I'm searching for my dad." Tears blurred his vision. He had to blink twice to get them cleared. When he got focus, he found a large pair of eyes staring at him. The Nitonian blinked twice then continued to stare. Was the creature mocking him? He deliberately avoided Landing as he pointed to the aerial. "Where's my dad?"

The Nitonian raised his tentacle again and touched Landing.

Fanar closed his eyes and sighed. This was going nowhere. First his mom and now his dad. How could this Nitonian do this to him? How could he take away what was left of his family? What had his father really done to him? Nothing. All his dad wanted to do was to help humanity find something better! He clenched his fists. He opened his eyes to see the creature staring at him. He ground his teeth as hot tears streamed down his cheeks.

"Why?" Like an explosion, the question came from Fanar with such an unexpected force that the creature jerked his entire body back. At that moment, a shot echoed in the meeting hall, followed by several more.

# CHAPTER TWENTY

## Outbreak

*Log entry for Contessa. Landing is in an uproar. That thing is still alive after being shot, and Dr. Huff is trying to make sure it stays alive, but several other people are saying we should let it die because it most likely killed Fanar's dad. And Fanar, poor Fanar is beside himself. One moment I think he's going to cry and the next he's angry. I wish he would let me help him in some way. (System mark: Day 53)*

**"W**here's Miss Shaw?" blurted one of the students.

Contessa glanced at the door. For the past couple of weeks, she had been helping Julie Shaw teach because Contessa's mom had insisted that she find a way to get out of the home and help the community. When visiting Contessa's mom, Julie had mentioned in passing that she could use some help. After Contessa had healed, her mother had volunteered her. It proved difficult to leave her mom for any length of time. Perhaps her mother had been right and Contessa relied on taking care of her mom as much as her mom had become reliant on her.

"I don't know. We'll find out after class," Contessa answered.

The school had been scheduled as the first to connect with the Oracle, but those plans had been pushed back because the materials hadn't arrived when the portal had closed early.

Each student had a desk with a display that allowed them to learn at their own pace. Because the ages ranged from five to ten, this was especially important. Contessa and Julie would clarify and guide students in their lessons and, more importantly, keep them on task. It surprised Contessa how difficult it was to have them focus for twenty minutes before the students would have a fun break. Had she been easily distracted as a child?

When lessons were complete and the students left, Contessa made sure the classroom was in order and trekked up I Street to Julie's house. The blinds were closed. She knocked on the door. After waiting for fifteen seconds, she knocked again. Getting no response, she was turning to leave when she heard a thump against the door. "Julie?" She pressed her ear against the door. "Julie?" She scanned the area. It didn't feel right to open the door, but what if she were hurt? "Julie, I'm coming inside."

She opened the door several centimeters and met resistance. Immediately she thought of the dead technician at the departure site after the explosion. Her breath shook as fear gripped her. "Julie!"

A grunt followed by a murmuring came from behind the door. Contessa pushed the door open enough to squeeze her head through. Behind the door lay Julie, her legs pressing against the door with the rest of her lying obliquely. With a little more force, Contessa moved the door enough to slip inside. She didn't see any blood, thank God, but Julie lay motionless. She rolled the teacher away from the door and onto her back before shaking her shoulder. "Julie! Julie!"

Julie's eyes fluttered open, and her lips moved. She swallowed. Her mouth moved soundlessly again, and her eyes closed.

*Is she dead? Her face is so pale! No, she's breathing.* Julie's chest rose in quick succession as she took in shallow breaths. *What should I do?* "I'm going to get Dr. Huff." She left the door open and ran down the street past the school. Her sides ached, so she slowed to a trot as she crossed the open area between the school and the medical clinic. She held her side and urged herself around to the entrance and went inside.

"Oh, my! Are you okay?" Susan, the nurse practitioner, rushed forward and took Contessa's arm to help her to a seat.

She waved Susan off. "I'll be okay"—she looked around—"where's Dr. Huff? I need Dr. Huff!"

"She's not here. I can help; what's the problem?"

Susan had been by their place to check on her mom only once, so Contessa didn't know her very well. Still, she had to get someone. "It's Julie Shaw, the teacher. There's something wrong with her. She didn't show up to class today, and when I went to her house, I found her passed out on the floor."

Susan talked as she hurried into her office. "Did you see any blood or bruises? How's her breathing? Pulse? What did her skin look like? Did you see anything nearby that might indicate if she had been eating?" She came out carrying her medical bag.

"No, I didn't see any blood. She was breathing very fast." What else had Susan asked? The questions had come so fast.

"What about food? Did you see any food near her?"

Contessa thought about what she'd seen. Had she seen food? She followed Susan out the door. "I don't know if she was eating anything."

Susan stopped and turned. "Which house?"

"I-4." Her side had stopped hurting, so Contessa could keep up with Susan, who briskly made her way to the house. Inside, they found Julie the way Contessa had left her.

Susan placed her fingers on Miss Shaw's wrist. "She has a pulse, but it's weak." She put on some protective gloves before she dropped to her knee and took a short tube out of her bag, which she placed onto Julie's finger. She then unbuttoned the shirt and took out a strip of thick contoured fabric with a silky black surface. She peeled something from the underside and pressed the cloth onto Julie's chest, covering part of the sternum and her heart. Immediately, the cloth showed four white cardio patterns moving across its surface. While the EKG took place, she pulled the tube off the finger and looked at the display. She studied the EKG reading and frowned.

"What? What's wrong?"

"I don't know." Susan watched the readout as it crossed the surface of the fabric. "I thought it might be a myocardial infarction. Unusual for someone as young as her, but not —"

"A what?"

"Sorry. A heart attack, but her blood is well oxygenated and this readout doesn't give an indication of a heart failure. Do you know who else lives here?" She peeled the cloth off of Julie and dropped it into her bag before doing a more thorough exam.

"I don't know their names, some scientists. They work in the lab at the end of the street."

"Mmhmm. When did you last see her?"

"At the school yesterday. Sometimes we talk afterward, but I wanted to check on my mom, so I went home. She seemed fine and was talking to one of the soldiers."

Susan bit her bottom lip. "I think Dr. Huff would want a blood sample, and we need to get her on an IV. Let's get her up onto the couch." After moving Julie to the couch, Susan faced Contessa. "I can draw blood here, but I want to move her to the clinic. Can you see if a rover's available?"

Contessa nodded and headed to the maintenance building. Fanar sat on the ground next to the entrance. He looked up, his dark eyes meeting hers. They weren't bloodshot from crying, but he didn't seem angry either. She couldn't discern his mood. "Hi, Fanar. I didn't expect to see you here."

His eyes shifted in the direction of the meeting hall. "I had to get away and do something." He returned his gaze to her. "I thought I would help out here, but it's locked up." He remained silent for several seconds. She was opening her mouth to say something when he continued, "He's dead, isn't he?" He wiped his eyes.

She wanted to hold him, but instead placed her hand on his. It wasn't right for him to lose his dad like this. Now he had nobody. "I'm sorry. I'm sure we'll find him."

"I wish… I wish I could have prevented it. Why my dad?"

"There's no way you could have known. From what I gather, nobody knew about these things being here."

He sighed and pulled his hand out from under hers to wipe another tear. "What are you doing here?"

*He acts a lot like Rhona, afraid to show weakness.* She pushed the thought aside. Julie needed her immediate attention. "Do we have any rovers? We need to get Julie Shaw, the teacher, to the clinic. I found her collapsed at her home. She looks really bad."

His eyes went wide. "I think there's one by the meeting hall." He waved for her to follow. "What happened?"

She told him about Julie and why they needed the ride as they jogged across the grassy area between buildings. "We're need to move Julie to the clinic."

Several people were on the road, probably to see that thing in the meeting hall. As Fanar had suggested, a rover sat in front of the hall, and Kellach approached it with a bottle in his hand. Fanar ran up to him. Contessa kept her distance while they talked.

"No, because this is rover four," Kellach said. He didn't wait for Fanar to respond before opening the door. "In all the commotion, we forgot to properly clean it after hauling the beast back here. Me, I'll get it properly cleaned." He held up the bottle in his hand. She didn't know what it was; maybe Fanar knew. Kellach climbed into the driver seat.

"Wait!" she yelled, but it was too late. Kellach turned the vehicle around and drove toward the maintenance building. What were they going to do now?

"Maintenance isn't that far. I'll run over and meet him and see if there's another rover," Fanar said. Without waiting for a reply, he ran up across the grass. Oswin might poke fun at him for his lack of running ability, but to her, he had become a solid runner.

Contessa sighed. Pretty soon, she'd be running as much as Fanar and Oswin. She loped behind Fanar. He had run off quickly, so he should be back with another rover before she got far. She passed the place of worship, expecting to see him returning in another rover. *Where is he?* Julie was in bad shape and time was critical. When she finally got inside the maintenance building, she saw only one rover and no sign of Fanar.

Where did he go? Where's Kellach?

Fanar came running from the front. "Sorry. I had to get the security code from the front desk. It got changed."

Contessa got into the front with Fanar. "She's in I-4. Hurry." She held on to the handle above the door. Fanar didn't disappoint and careened onto the road.

In only took a minute to get to the house. She leaped out and ran into the house and stopped. Her hand went to her mouth. Julie's face had become ashen.

Susan acknowledged them entering the house. "I tried everything."

After taking Julie's body to the clinic, Fanar drove Contessa home and stayed for an hour. The hug he gave before leaving was comforting; she wished she could have held him for another hour. When she got home, the place was quiet. Rhona snored as Contessa climbed into bed. How was she going to break the news? Julie's pale face dominated her mind. She should have run faster or done something. She had been too late, and Julie had died.

Contessa woke to an airhorn. The horn blasted for five seconds with a ten-second break followed by another five-second blast. The blast repeated another three times before it stopped.

"What's that?" Rhona asked.

"I don't know. Let's find out." They got dressed and ran out to find their mom coming out of her room. "Mama, what was that?"

Their mother's eyes appeared tired. "Let's go see. Sounds like it came from the meeting hall."

They found the street crowded with people walking toward the meeting hall. Contessa told Rhona and her mom what had happened. Others crowded around her to listen and ask questions. When they arrived, three of the soldiers stood at the entrance behind three of the board members. They weren't letting anyone inside.

Dr. Yamamoto raised her hand and waited for silence. "Everyone..." She closed her eyes and sighed before continuing in a loud voice. "As some of you may of heard, Julie Shaw, our teacher, died yesterday." She raised her hand again and waited for the murmuring to cease. "Dr. Huff"—her voice shook, and she took another breath—"was found dead this morning, seemingly from the same ailment that killed Julie." At this, the crowd grew agitated and questions were thrown out.

Contessa was so wrapped up in the death of Julie, she hadn't realized Dr. Huff wasn't standing with the other board members. Dr. Yamamoto raised her hand again, and the crowd grew silent. She wiped her eye and took a deep breath. "There are four people in the clinic who are sick. It has been brought to my attention that the rover used to transport the native creature had not been washed down properly after it arrived here. So we are asking everyone to check for signs of sickness and to do a thorough wash- down of themselves and anything that came in contact with the rover. We will also be restricting access to the meeting hall until further notice." She stepped away.

Are they saying that thing made everyone sick? If so, maybe it hadn't kidnapped Dr. Monier. Maybe he'd gotten sick and died. And how am I still alive?

# CHAPTER TWENTY-ONE

## Visiting Hours

**W**hen she was six, Contessa had gone to her first funeral. She remembered because she had asked her dad why the big box made people cry. When her dad asked if she wanted to see, she shook her head. If it made people cry, she didn't want to see. Tonight there were no boxes, only images glowing in the air from tiny projectors. Images of Dr. James Monier, Dr. Christine Huff, Julie Shaw, Randy, and four others she had seen but didn't know their names. As her eyes paused on the images of Julie and Randy, her head felt like it was floating above her numb body. *How could this happen?*

A few hours after the announcement at the meeting hall of the deaths, those at the clinic had died. Two others had also been found dead.

Everyone in Landing stood in the field where they'd first arrived.

Everyone except Fanar. She didn't know his whereabouts.

Like so long ago, her dad stood next to her. She intertwined her arm with his and leaned on his shoulder as tears streamed down her cheeks. No big box needed. Rhona held on to her father's other arm while their mother rubbed Rhona's back. Kellach stood near the floating images and spoke of Randy's dedication to making people feel safe and how he'd been instrumental in finding the creature. He said a lot more, but it was difficult to hear.

"The creature that killed him," Oswin whispered. His mom held her finger to her lips, silencing Oswin.

He didn't have to say it. Contessa was sure all of Landing was thinking it. Randy had been instrumental in finding and capturing the creature that killed him. Dr. Huff had died after saving the creature's life. People surmised it had left some toxic residue in the rover and those who had used the back area had been affected. Their bodies were in storage so they could be interred on Earth when the portal opened in five days. She craned her neck and saw Fanar at the back of the gathering. He stood transfixed by the image of his dad shimmering in the air alongside the others. She let go of her father and walked over to Fanar.

"I'm sorry," she said.

His eyes met hers. "He isn't dead. We haven't found him. He isn't dead." *Denial. Isn't that the first stage of grief?* She wanted to hold him and let him know it would be okay. She studied his mussed hair and long face. He needed some sleep. She put her arm around him and faced the front. She saw Randy's older brother to the side. *What's his name? Mike. That's right.* Like Randy, he was fit and had beautiful brown eyes that were currently wet with tears. *It must be hard for him. At least Fanar has his friends. Mike has nobody now.*

*This is Fanar. We had a memorial a few days ago for eight people who died, including my dad. Everyone tells me how sorry they are for my loss. Before the funeral I believed he had died, but now I don't think so. How can he be dead if we never found his body? I don't know. I still feel the answer is with the captured Nitonian. (System mark: Day 56)*

"Why do you want to do that?" Oswin asked before shoveling another bite of his breakfast into his mouth.

Fanar shifted the contents on his plate around. "I think there's still a chance to learn something."

Oswin shook his head. "It's caused nothing but trouble since we got here. First it sabotages our stuff, then it tries to kidnap Contessa, then it kills people? No, we don't need to go visit. Besides, they aren't letting anyone inside."

"He didn't mean to kill anyone."

"How can you say that? There are eight people dead, including your dad! Honestly, I don't know how you can defend it." Oswin picked up his empty plate and went into the kitchen before heading toward the front door. "I have work to do."

Fanar watched Oswin leave. *What's his problem? Can't he see the Nitonian holds the answers?* He tossed his fork onto the plate. He didn't feel like eating, even if it was something Rhona had put together for them. He picked up his plate and scraped the food into the compost collector. To get answers meant spending more time with the Nitonian, and to do that, he would have to convince the soldiers he should be allowed to see him. He walked across the street to the meeting hall and went inside.

One of the two soldiers stood. "You can't be in here."

"I helped bring him in and I didn't get sick," Fanar said.

"Doesn't matter. We were told to not let anyone near it while they decide what to do." The soldier's frown deepened. "I don't understand how it's still alive."

"Dr. Huff said the thick skin has a high mineral content, so the bullets didn't fully penetrate or got deflected."

The soldier's face twisted in anger. "Not that! How are we allowing it to live? He killed my brother! We should kill it just for that."

*His brother?* Fanar realized he was talking to Mike. *No wonder he's so angry.* Fanar looked across the chamber into the conference room. The creature had inflated himself and resembled a boulder again. *This must be how they sleep.* He said to Mike, "If we're going to find my dad, we need to get some answers. Just let me try to communicate with him. I was making some progress the other day before…" His voice trailed off. If he hadn't shouted, then the creature wouldn't have reacted the way he had and gotten shot.

"We're under orders from the board—what's left of it anyway. Get oneof them to agree and you can talk to it. Just keep me away from it."

Fanar nodded and went outside. The board members had worked on this project together for the past several years. Convincing one of them wouldn't be an easy task. Dr. Yamamoto always had a soft spot for him; he would talk to her first. He strode to the food processing and storage building next door and went inside. Nobody. They were probably out gathering more food. She was the one he really wanted to talk to, and she would most likely be gone for a couple of days. She had once told him while others gathered food, she searched for plants to experiment with crossbreeding. Finding fast-growing species and combining them with the tastier varieties would be highly beneficial. Interesting, but it didn't help him now.

He wandered up the road, wondering where another board member might be. *Maybe I should check the labs at the end of each street.* As he contemplated this, Kellach drove up from behind him and stopped.

"Just the man I was looking for," Kellach said. "Hop in, I need your help."

"Does it have to be now?"

"We need to run some cycle tests on the portal generator so if something isn't working right, we can get the parts when Earth opens the portal."

Fanar weighed this in his mind. He really wanted to get in with the Nitonian, but making sure the portal generator was ready for use was critical so contact with Earth was there when needed. "How long will it take?"

Kellach bobbed his head back and forth a few times as he considered the work. "Two hours, give or take."

With a sigh, Fanar climbed into the rover. They drove to the building that housed the portal generator and went inside. The equipment was identical to that on Earth, but smaller. "I thought you tested this after we fixed the water bladders."

"I just made sure there wasn't any obvious damage. Today we do some actual testing." Kellach approached the first cabinet and tapped a button, bringing the display to life. "Running these tests will add another two days to the collection time, but it's necessary. First thing we'll do is check the energy output from storage. Now watch." He tapped the panel to enter test mode and then went to the battery test area and tapped again.

A surge of energy pulsed through Fanar's body when Kellach tapped the last button. He shivered.

Kellach scowled at him. "You okay?"

"Yeah. It tingles."

"Hmmm, you must be extra sensitive. This is the lowest setting for tests, just a few hundredths of a percent. Enough to get a reading. Never heard of anyone feeling anything. Me, I didn't feel a thing. You sure you're okay?"

"Yeah, it didn't hurt or anything. Just… weird."

Kellach stared at him for a second then directed Fanar's attention to the display. "See how we're at the ninety-five percent mark? That's perfect. We don't want to go over ninety-eight percent. Next is the signal cycle test." He tapped some more buttons and, after waiting nearly thirty seconds, a green waveform was shown. "It will be green or red."

Fanar felt another pulse of energy just as the waveform was shown. This would take some getting used to. "And red is bad."

"You got it. I'll start on the other side and meet you in the middle."
"Since it takes so long, can't we start several of them at the same time?"
"No, the test wouldn't work right. They have to be done individually along each bank." He went to the end of the long cabinet and disappeared from sight.

Fanar realized the long cabinet contained twelve identical parts. Each would need to be tested. He went to the end and counted. Forty cabinets—that was a lot of testing. He went back to the second station and started the tests. If these were the lowest test settings, then his dad must have been using much higher test settings back on Earth before they departed. No wonder that technician had been concerned about energy use. As Fanar moved from panel to panel, he likened the energy jolt to that of a small static shock. Surprising, but harmless. It wasn't difficult work, just time consuming. If anything went wrong, he wouldn't have a clue how to fix it. Eventually, he and Kellach met in the middle and finished the testing.

"No problems?" Kellach asked.

"Nope. All green. There were some stations that showed ninety-six or ninety-seven percent on the energy output, but nothing beyond that."

"Good. The bladders are almost full for Earth."

"You need me anymore? I have something I want to do."

"Not now, but I do need you tomorrow at the G lab. They want us to verify their power output while they're out."

Fanar nodded and left. *There goes two hours I can never get back. Now to find a board member and convince him I should be allowed to visit the Nitonian.* He walked down to the I lab. It was empty. He proceeded across the plum-colored grass to the H lab. Like most of the other labs, the large room had several bench desks in the middle and long workbenches along the sides. A few doors led to smaller rooms.

One of the scientists noticed Fanar enter. "Can I help you?"

"I'm just looking for one of the board members. You know where one might be?"

"You could try the B or C labs. One should be there."

"Thanks," Fanar replied. He sprinted down the street toward the main circle. He didn't want to waste any more time. As he neared the road, he saw someone in the middle of the main circle. As he got close, he recognized the stout figure of Dr. Chernoff holding something near his face. Being the resident entomologist, probably an insect. Fanar approached him. Dr.

Chernoff jerked his bald head up to reveal a magnifying eyepiece strapped to his face, his bushy eyebrows peeking over the top. "Fanar. You startled me." He rolled the final R in Fanar's name with his thick Russian accent. He peered at the insect held by the tweezers in his hand and smiled.

"Good, I didn't drop it. A most amazing find. Right here in Landing!"

"Dr. Chernoff, I need to ask—"

"Too bad it isn't alive. I've searched and can't find any other like it. I suspect it came back on a rover from a different area."

Fanar had learned some time ago that scientists lived to share their discoveries, and there was little chance for him to get his question in until he indulged the man. He quietly took in a deep breath and opened his eyes wider. "What's it look like?" *Like I care?* He looked at the tiny insect thrust in front of him. Nearly a centimeter long, it had a narrow head and swollen brown body surrounded by a dozen legs and a visible stinger. Fanar scowled. "I've seen this before."

Dr. Chernoff's eyes went wide. "You must tell me. Where did you see it?"

*Where indeed.* He stared at the insect, trying to remember—oh, yes! "By the river, past the falls. It fell onto my hand when we found Contessa."

Dr. Chernoff marched toward the maintenance building. "You must show me."

"I can't. I need to see the Nitonian."

Dr. Chernoff turned to face him.

"Please. I really need to talk with him. I still think I can get through to him."

"Fanar, you were at the memorial. Your own father—"

"He isn't dead. He can't be. We never found his body."

Dr. Chernoff shook his head. "We put the creature in isolation because of the toxin. We need —"

"But I'm not sick. I've been around it at least a couple of times and I've never been better. Maybe I'm immune. Maybe it only works on some people. I'm not sick. I was one of those who carried it onto and off the rover. If it were going to affect me, wouldn't I have gotten sick and died too? I really believe he holds the answer to where my dad is, and I won't even touch him." Fanar waited and agonized. *If I can't get permission here, then I will go elsewhere.* He couldn't guarantee either of the last board members was currently at Landing. Coming to Niton was not only a resource- gathering mission, but a scientific expedition, and that meant the scientists had to spend time away from Landing. He watched and waited as Dr. Chernoff considered his argument.

Dr. Chernoff inhaled and gave a slight nod.

"Thank you! You won't regret it."

"I already do. My direct verbal consent is needed, so I must go with you."

# CHAPTER TWENTY-TWO

## Escape

**F**anar stood inside the meeting room's closed door, studying the Nitonian. *What is this creature thinking?* A sense of sadness and despair washed over him as he contemplated how to proceed. *If I'm sad, I can only imagine this Nitonian feeling the same, locked in a room against his will.* Did they offer him food or water? Yes, in the corner he saw an empty plate and bowl. *I wonder what they gave him. Whatever it was, he looks different.* The creature's brown coloring had faded. Fanar glanced at the wall display. It showed Landing and the surrounding area with the image of his dad in the upper corner. Would he be able to use that or should he take a different approach? He suddenly realized no real intelligence had been displayed. What if this creature had just been foraging for food when it caused all the damage? What if all this time he was simply working on instinct?

*Of course, he showed something beyond instinct when he pointed to the image, but even a cat could do that.* In the meeting hall, the soldiers were watching but weren't holding their guns. In fact, they hardly seemed to be interested in what was happening at all. Randy's brother, Mike, shot a glance their way then resumed talking with the other soldier.

Perhaps Fanar had been overthinking this all along. His dad had tried to tell him before he disappeared; there were no signs of civilization. No intelligent life. *So where did Dad go?* None of the rovers were missing. He played with the uplink on his wrist. It had previously belonged to one of those who died.

"Shanar," the creature whispered.

The hair on Fanar's arms stood upright. He held his breath, afraid to breathe. He caught slight movement in his peripheral vision. He moved his head the minimal amount to bring the Nitonian into view.

The creature repeated Fanar's name in its odd fashion with a sort of whistling sh sound in place of the F. Its eyes intently watched Fanar.

Fanar touched his chest with his finger. "Fanar. That's me." He glanced out the window. The soldiers must not have heard anything because they had moved toward the building entrance and were idly talking with each other. *Should I call them over?* He looked at the creature watching him. No, everything seemed okay. Maybe the creature did know where his dad was. Fanar walked to the wall display, keeping his eyes on the creature as he moved. He touched his chest with his finger and repeated his name, then he pointed to the image of his dad. "James." He did this a few times then pointed to the creature and waited.

"Ciabgan."

Fanar tried to repeat the name. "See-ob-gun?" He pointed to himself and gave his name then pointed to Ciabgan and again tried to say the name. "Ciabgan." Slowly the creature gave his name. The initial portion wasmore of a whistle than a normal phoneme, and the G was guttural.

Fanar barked a short laugh. *Communication!* He ran his fingers through his hair. *I was right all along! This changes everything.* He pointed to the image of his dad and, with his finger, drew an imaginary circle around the area map. He watched Ciabgan to see his reaction. Then he pointed to various random points and stood back. "Where? Where's James?"

Ciabgan watched Fanar and sidled up to the display. With a tentacle, he pointed to James. "James." Again, the S was a whistling sound.

Fanar nodded. "Yes, James. Where's James?"

Ciabgan pointed at landing. "Landing." It started to speak in its native language of whistles and tongue clicks along with other sounds.

Fanar closed his eyes. *So much for communication.* Ciabgan was just pointing out the major landmark on the map. He was dismayed and… something. He wasn't sure how he felt. He put his hand over his eyes and rubbed his temples. As he rubbed, he heard a familiar crackling sound, and a flash penetrated his eyelids. He opened his eyes just as the door flew open. In less than a second, the soldiers were thrown back and the outside door exploded open. Nothing of Ciabgan but the familiar odor was left behind.

Rhona reporting in. I'm glad Dr. Yamamoto let me come with her on this food- gathering trip. For the last few days, I've been helping gather food and learning a lot about the various plants that were discovered in the initial survey. She says I'm great at identifying the edible plants and creating new ways to prepare them. Of course I am; it's food! I can't believe how much we have to gather to feed everyone. Guess that's why we haul these giant trailers behind the rovers. Had some crazy dreams last night, but I think it's because I'm in a strange place.

We're returning to Landing today. (System mark: Day 57)

Rhona slapped her uplink to dismiss the activity, glad she'd decided to wear it on this trip. When they got to Landing, she would sync it with the Oracle. A portion of the trip was to not only gather food, but to start cultivating the land. Dr. Yamamoto had explained to Rhona that small cultivations in various areas should be safe until they got the materials from Earth for plant towers. Vertical greenhouse towers that would rotate and focus the sunlight onto the plants, grown from seeds that Earth would provide, in an optimal way while keeping them watered.

"That's your log entry?" her tentmate, Tammy, asked. Tammy was nice. When Dr. Yamamoto concentrated on other tasks, Tammy took charge. Rhona felt lucky to have her as a tentmate.

Rhona scowled. *What's wrong with my log entry?* "Yeah, why?" "It's so upbeat. I can't say mine are."

"I just think it's exciting. Discovering new things."

"Maybe I'm just too focused on all that's gone wrong. Help me pack the tent?"

Bright lamps lit the campsite as they packed. The trip to the cave with the boys had been more helpful than expected. Not only did she know how to put away the tent, but use of the waste management packs didn't have to be explained to her. When she asked why they had to be used so far away from the settlement, she learned human waste contaminated food sources.

"We leave in two hours," Tammy said.

"Is that enough time for me to go look at something?"

"Um… I guess? Where you going?"

"By that hill"—Rhona pointed to a hill a few hundred meters behind herself; the focus of their energies had been away from the hill, so nobody else had been around it—"I saw some orange berry bushes. I just want to get some to see if they're edible."

Tammy glanced at the hill. "Yeah, that should be okay." As Rhona started to leave, Tammy raised her voice. "Grab protective gloves and a bag and remember to get samples of the entire plant. Fruit, leaves, and branches. If possible, some of the root too."

"Right!" A squeak of delight erupted from Rhona's mouth as she gathered the equipment. *Another discovery!* First it had been the cave, and now it could be something that would benefit everyone. The plants everyone ate were okay, but she wanted something sweeter. Just before the memorial service the other day, she had eaten the last of her chocolate bars, and now she wanted, no needed, something to satisfy her sweet tooth. Chocolate was her first choice, but who knew how long such a low priority item from Earth would take?

She strapped on the hand light, bounced past the hill, and stopped. Behind the hill were several smaller ones, their dark outlines barely discernible. A slight omission, but if she could find the berries quickly, then nobody needed to know. Now where had she seen the berries? She advanced while moving the light across the rocky landscape near each of the hills. The area behind the initial hill looked dramatically different from where they had been working. Long plum-colored grasses grew in clumps around and on the craggy hills with the occasional tree or bush, but no orange- colored berries could be seen. Had they been eaten? Up to now, besides the Nitonian, those little mousy things were the biggest creatures she had seen. She had once overheard some people talking about how the area might have been through a recent ecological disaster and that would explain the lack of larger animals. If so, larger animals would eventually make their way back. Rhona approached the nearest hill. *I know this is where I saw them.* She walked along the base, careful to not trip on any of the rocks scattered around until she had traveled well past a quarter of the way around it. Nothing. She faced the next hill and started around it in the other direction. Her light came across a bush, but no berries were to be seen. *Is this even the right kind of bush?* She had seen it at a distance when it was bright outside.

Now, in the dark, she wasn't sure. A quick glance at her uplink told her little time remained. She'd better find those berries soon!

She turned to go toward another hill and stood still. Which one had she just looked around? The one to the left or right? Her eyes darted between her options. She sucked in a breath of encouragement and started toward the one on the right then shifted to the one on the left. This had to be the one closer to where she'd started. As she rounded the hill, the beam of her light exposed a large grey boulder. *Another Nitonian!* It was much bigger than the one they had captured. *How did I miss it before?* She lowered her light. Her legs anchored themselves to the ground, refusing escape. Her pounding heart echoed in her ears. Fanar might be unreasonably fascinated with these, but all the deaths at Landing had taught her an important lesson. Stay away! She had to get past it so she could get back to the others. No movement.

*Maybe it's asleep and I can sneak past?* Like trudging through wet cement, her legs grudgingly accepted the commands to move in a wide arc around the slumbering beast. Her eyes remained transfixed on the grey surface as she inched forward. Each second was a second too long. She placed her foot forward, and a hole swallowed the bottom portion of her leg. Before she could think, she screamed! She scrambled to get up and swung her light toward the creature, her eyes wide. Still no movement from the grey mass to her left.

She held her breath, her heart pounding a hundred and thirty beats per minute in her ears. *Is it alive?* All the noise she'd made, certainly it would have woken up. She glanced to the right then back. She could continue on, but curiosity of the large mass and a desire to leave warred inside her. Curiosity won. She tiptoed forward with the light fixed on it until she stood a few meters away and frowned. This wasn't a Nitonian; this was actually a boulder that resembled a Nitonian sitting in the rocky area. She stepped forward and held her fisted hand over it for a few seconds then tapped. Her knuckles hit a rocky surface, not the leathery shell like that one they'd captured. She sighed in relief, then her eyes went wide as she scanned the area and her new dilemma made itself painfully known. *Where am I?*

Her light bobbed in front of her with tears streaking down her cheeks as she stumbled across the landscape toward the last hill she had searched. At least she hoped it was. *This is just like the caves. How could something so simple go so wrong?* The hills weren't that big, only twenty meters or so in height, and there were only a dozen spread across the area.

She spun around, not sure which way to go. Right? Left? She sat and buried her face into her hands, sobbing. She didn't care. Her obsession with being the one to find something new had driven her into doing something stupid.

"Rhona?" Tammy called out. Her voice came from a short distance away.

Rhona lifted her head. "Tammy?" "Rhona, don't move! Where are you?"

"I'm here! I'm over here!" She stood and wiped her eyes.

Tammy's light blinded Rhona as she came around the bend. "There you are. Did you find your berries?"

"No, I think they got eaten." She ran up to Tammy and followed her past two hills, bringing them to the hill Rhona had pointed out just under an hour ago. She felt so stupid when she realized how close she was. Maybe she wasn't cut out for exploring.

Before they went around the final hill, Tammy stopped and faced her. She brought her light up just below Rhona's face. "Sorry you didn't find what you were looking for." She held out a small towel. "You might want to wipe your eyes. I think in the future it would be best if someone else went with you, okay?"

Rhona nodded and used the offered towel. Right now, she was glad there would be a next time.

After returning to Landing, Rhona transported, sprayed, and stored the gathered items until she thought she would drop from exhaustion.

"You've earned your credits today. They'll go a long way to offset your costs," Tammy said.

"My costs?" *What is she talking about?*

"Sure. Everyone sixteen and older is expected to pay for their food, maintenance, etcetera. Earn their keep. Food collection pays well. Didn't you know?"

Rhona shook her head. No, she hadn't been told, but thinking back, it made sense. Kellach had said they had to pay to use the rover. Maybe he'd let it go the first time, but she was pretty sure Fanar had needed to pay the other times. A thought struck her. "Wait, I've been paying for my own food?"

"Yep. Your mom's probably been dealing with it. See you tomorrow? I can hardly wait to see what new ideas you come up with on preparing the food."

Rhona grinned. It was nice to be appreciated for something she thought was fun anyway. After her recent experience, maybe she should limit her explorations to food dishes.

Instead of walking on the main road toward her house, she went the other direction. It had been a long day, and it would be nice to spend some time with the boys. Contessa was probably there anyway.

She wasn't disappointed; Contessa was the one to open the door. Inside, she found Oswin arguing with Fanar.

"I think maybe you let it go," Oswin said. His eyes narrowed. "How else do you explain it getting away so easily?"

Fanar's face glowed red, and his nostrils flared. In a growl, he said, "I did not let him go."

"What's going on?" Rhona asked Contessa.

"That thing escaped." She paused. "Fanar says it can talk."

Rhona's eyes went wide. "What did it say?"

"He said Fanar's name and gave his own. Sheojen?"

"It's Ciabgan," Fanar said. He turned back to Oswin. "How could I have stopped him?"

Contessa continued, "Anyway, Fanar tried to get it to tell him where his dad is, and it just pointed to Landing on the map." She glanced at Fanar. "He has a fascination with this thing that goes beyond reason."

"Fascination beyond reason," Oswin said. "Thank you, Contessa. That's exactly what this is. We shouldn't be trying to talk to it; we should be figuring out how we can protect ourselves from it."

Rhona nodded. "I thought I saw one of those things this morning and almost peed my pants. What are we doing to keep ourselves safe?"

"Don't you see?" Fanar pleaded. "He was just trying to get away. We don't have anything to worry about."

Rhona couldn't believe her ears. *Nothing to worry about?*

Oswin must have been reading her mind. "What do you mean, nothing to worry about? Don't you remember how much equipment got destroyed? What about all the deaths? Your own dad?" He shook his head. "I don't get you." He stormed past Rhona and left the house.

She followed him out the door; maybe she could get some answers from him. "I don't understand. What's Fanar doing?"

He spun to face her and jerked his hand up. "He—" He gritted his teeth. "He thinks we should go after it. Like somehow we're going to all be friends and somehow it will tell us where his dad is."

"But his dad is dead. That thing killed him. Surely he knows that."

Contessa came outside. "He's not talking to me. The other day he seemed to be accepting his dad's death, but now that the creature what's-his-name—"

"Ciabgan." Oswin said.

"—Ciabgan is talking, he seems to think something else. I don't know." Fanar came out of the house carrying a pack and stormed past them. "Where you going?" Contessa called after him.

"To get answers," he called back.

"Maybe he should talk to Dr. Williams," Rhona said. "He needs some help."

# CHAPTER TWENTY-THREE

## Crystal Cave

*This is Fanar. I have two days to find my dad before the portal opens. When I do,*

*I'm going to convince him to come back to Earth with me. (System mark: Day 57)*

**F**anar strode along the road toward the food processing building. Peeking out from the back corner of the building sat a rover. He had intended on checking one out from the maintenance building, but this one was closer. Not seeing anyone around, he went over to it and disconnected the trailer. The swish of the building's back door startled him, and he ducked behind the side of the rover.

"I'll hose these down then get them to maintenance."

He didn't recognize the voice of the woman, but he knew what he had to do. He reached his hand up and pulled the door handle. Slowly he opened the door and pushed his pack ahead of himself into the cabin of the vehicle. He quietly closed the door. Keeping his head low, he tapped the console and entered his destination. The waterfall.

"What—?" The rest was lost as the rover accelerated.

Fanar popped his head up just enough to see Dr. Yamamoto and someone else staring at the seemingly empty vehicle race away. The vehicle headed north up the road and cleared the buildings before turning east to circle the settlement. Once he was away from Landing, he lay back. The trip would take some time, and he could use some sleep. The waterfall called to him in a way he couldn't explain.

After some time, he realized it wasn't the waterfall that captivated him, but the cave behind it. He was being drawn to it like a bird back to its hatching nest. It was instinctual, beyond explanation.

The sudden lack of motion woke him. The thundering pulsations of the waterfall vibrated the shell of the rover. Two of the moons were visible, but it wasn't enough light to navigate, so he switched on the rover lights. He followed the path he and the others had taken before and crossed the river. When he came to the path behind the waterfall, it was awash with moving shadows as the moonlight poured through the falling water. The movement of moonlight and shadow made him see what he had subconsciously picked up before. The texture of the path and the cliff face had a repeating pattern made to appear natural. Someone had made this path by cutting into the cliff face itself. He rushed up the path and into the cave. He pulled a hand light out of his pack, strapped it to his wrist, and scrutinized the walls. He wasn't an expert, but his initial observation marking this as a lava tube still seemed accurate.

He strode through the tube to the large cavern. This time, he'd come prepared. He reached into his pack and pulled out an electronic pad and a domed object about nine centimeters in diameter across the base. After placing the dome onto the floor and connecting it to the panel with a wire, he tapped the panel. The dome measured and mapped the space using a couple thousand tiny unseen lasers. He then moved to the other side of the dome and repeated the process. The second time removed him as a measured object. He unplugged each side of the wire and stowed it along with the dome into his pack before studying the resulting image shown on the pad.

The number of outgoing tunnels surprised him. Based on the image, there were more than ten tunnels of various sizes leading away from the cavern. He turned around to face away from where he'd come in. A large tunnel should be on the other side of the chamber. He looked again at the image displayed on the panel and thought about Rhona getting lost and how they'd had a difficult time finding a way out. *Should I be risking this?*

If it meant getting his dad back, then yes. He walked forward. When he arrived on the other side, he found the tunnel. It seemed to be another lava tube but wasn't as big as he'd thought it would be. Just over three meters wide. He walked for thirty minutes through it until he came to a fork. He recorded his decision to go left on his newly acquired uplink and continued.

Every ten minutes, he would stop and map the tunnel. After another hour, he came to another chamber. He mapped it and had a snack.

*What am I doing here? This is probably a waste of time, wandering through empty tunnels in a hopeless attempt to find my dad.* A part of his brain whispered, *My dead father.* Most likely under a bush somewhere. He shook his head and picked up the pad and noticed only one tunnel on the other side of this chamber, about two and a half meters above the floor. A fact that had escaped him before eating.

Time ran short. He had to find his father and get him back to Landing before the portal opened. How was he going to get up there? After crossing the chamber, he reached up and couldn't touch the bottom of the tunnel entrance. He tossed his pack up into the tunnel and stood back. It was over his head, so he couldn't run and jump up into it, but maybe if he had a running start, he could jump and grab the bottom lip? He would try that. He adjusted the light and ran.

Fanar landed his foot against the wall about waist high and tried his best to propel himself upward. Instead, he pushed himself away from the wall and landed on his back. His breath exploded out of him, and his right arm and hand smacked the floor, causing the light to go out.

"Ow!" That had really hurt. He rubbed his arm. It didn't feel broken, thank God for that, but it pulsed in pain. He tapped the light switch. Nothing. He rapped the body of the light with his knuckles and then tapped the switch again. Still nothing. "Great!" he growled under his breath and cursed himself for his lack of ability.

His back and arm throbbed. He was not happy. *I should probably go back.* Would he be able to find the right tunnel to get back to the rover? He didn't have an extra light in his pack, but he could at least use the light from the electronic pad he was carrying to see a little. It would be enough. It could also be enough to continue a little further. He sighed, knowing either way he would have to retrieve the pack he had tossed up into the tunnel.

With a grunt, he got onto his feet. He was going to be sore for a while. He hoped he was facing the right direction. He moved forward with his hands out until he got to the wall. Faint white blobs of light dominated his sight. A phenomenon he'd experienced before when in a dark environment for a length of time. 'Phantom light,' his dad had once called it. Fanar squeezed his eyes shut for a few seconds then opened them. It helped.

The opening of the tunnel should be just above him. He held his hands lightly against the wall and jumped. The rough rock lightly scratched his fingertips. No opening. He stepped to his right and tried again. His right hand went into the empty space of the tunnel before he fell back down. He stepped to his right and jumped again. This time he grabbed the bottom of the opening. His arm screamed in protest, and he immediately dropped to the floor. *Maybe I broke it after all.* He gingerly squeezed along his forearm and winced when he got near his elbow. If he was lucky, it was only a bruised tendon. Either way, this would be difficult.

He jumped and grabbed the bottom of the opening again. Ignoring the protests from his arm and back, he hoisted himself up using what strength his arms could provide and a little help from his feet against the wall until he was chest high into the opening where he could arm crawl and pull himself fully in. *It's official. I'm crazy.* He collapsed, panting with exhaustion. Who knew a simple jump could take so much out of you? What he needed was a cold pack and a day of rest. For now, he would just take some rest.

Fanar woke with a start and winced. Sleep proved difficult to achieve due to his aching body, but he had finally fallen into a restless sleep for a little more than thirty minutes. He reached for his pack. The light from the pad should provide enough illumination for several hours. Enough time to explore the tunnel and make it back to the rover. He moved his hand around the area, searching for the pack. Where is it? He crawled to the side and bumped his head against the tunnel wall. "Ow!" He put his hand to his head and felt wetness. He slowly shook his head. This was not going well. He was sore and hungry.

If he could find the pack, he could at least get something to eat. He was also pretty sure he had packed something that would seal the cut on his head. Slowly, he moved back in the other direction, putting his hand out so he wouldn't bump into the other wall.

"How far did I throw it?" he mused aloud. After meandering his way deeper into the tunnel a couple of meters, he stopped. He definitely hadn't thrown it this far, which meant someone had taken it.

What did his dad like to say? *"Quand le vin est tiré, il faut le boire"*? Fanar hadn't really learned French, but he surmised it meant something about completing something you'd started. He sighed and started moving down the tunnel. He highly doubted he would find his pack but wanted to be sure. Something inside him said Ciabgan had taken it. He was probably crazy to think it, but it made sense to him right now. As he continued, the tunnel curved and angled downward, growing in size as it went down. Giving up on the pack, he stood and continued. A little further the tunnel curved again and came to a fork. He went left.

A quick glance at his uplink revealed he'd been moving for a couple of hours since his nap. Fanar sighed, agonizing at the slow progress. At each fork, he took the left so he could more easily find his way back. After a long while, the phantom light returned. He squeezed his eyes closed for several seconds then opened them again. He could still see it. With a scowl, he tried again. It had always worked before. When he opened his eyes, the light was still there. *Wonderful.* Not only did his arm throb with each movement forward and his back ache with each shift of his body, but now he had to deal with this.

The tunnel curved again. The light got brighter. *Am I imagining things?* He waved his hand in front of his face and saw it move. This was no phantom light! He increased his pace toward the brighter light. Another branch, but this time he went right. Soon the tunnel was lit enough he could run if his back weren't so sore. A sharp turn to the right and he stopped. Glints of green and blue mixed with white light washed over him as he stood at the entrance of a cavern lined with glowing crystals of all sizes.

Fanar stood mesmerized. The cavern was about thirty meters in diameter. He was so overtaken by the spectacle of shimmering light that he didn't notice the Nitonians until one stood near him. The movement caused him to jump back. Others moved in. They were much larger than Ciabgan. Nearly double the size. He stepped back and bumped into another one! He was surrounded! He swallowed. Were they going to kill him for what the colonists had done to Ciabgan?

One moved closer until he stood face-to-face with Fanar. His large red eyes with the thorn-like yellow dots stared at Fanar. Fanar's chest constricted as fear squeezed all the breath out of him. His hands shook. He shied away from the gaze, but the Nitonian shifted his head around to meet his own movement, and he couldn't escape it. Then his head exploded in pain.

# CHAPTER TWENTY-FOUR

## Boom!

*Oswin Carter's log entry. Last night, Fanar stole a rover. We believe he is in pursuit of that Nitonian, Ciabgan. The board has asked Contessa, Rhona and me to talk with them about his odd behavior. (System mark: Day 58)*

The darkness outside reflected Oswin's mood. His best friend had gone off the deep end, and now he had to talk to those in charge. How could he possibly defend Fanar's actions?

"Hey, it isn't your fault your friend's crazy," Rhona said. She put her hand on his shoulder. "It'll be okay."

"I just hope he's okay." He opened the door to the meeting hall for the girls then followed them inside.

Chairs were arranged in a circle in the middle of the gathering area. Dr. Aiko Yamamoto, Dr. Paul Williams, and Dr. Anton Chernoff, the three remaining board members, sat waiting. Dr. Yamamoto stood and smiled. "Good morning." She walked to Oswin and took his hand. "It will be fine. We're just concerned for Fanar."

He nodded and followed her back to the circle and sat. If he heard it was going to be okay one more time, he wasn't sure what he would do. He had just assumed that Fanar had stayed in one of the empty houses to cool off. He hadn't even known Fanar had taken the rover until earlier that morning. He felt responsible because he didn't follow Fanar. Maybe if he had stopped Fanar from going off on his own, he wouldn't be missing now.

Oswin had only talked to Dr. Yamamoto before now. Probably because she was a horticulturist and in charge of the food. Dr. Williams was a psychiatrist and Dr. Chernoff an entomologist. "Are we going to search for him?" Oswin asked.

"We have a couple of teams hunting for him now. Maybe he'll return on his own. He has the rover for shelter," Dr. Yamamoto said.

Oswin cleared his throat, not sure what to say. He glanced at the conference room where Ciabgan had been held.

"You look, because it can't be helped," Dr. Chernoff said. "Knowing the creature was in that room, I find myself looking too."

"Fanar spent a lot of time with it," Dr. Yamamoto said. "Before last night, Oswin, did you see any change in him?"

"Well, yeah," Oswin replied. "When his dad went missing. He was beside himself and wanted to find it. He barely slept and kept insisting that Ciabgan had his dad."

"Fanar was the first to know about this creature?" Dr. Williams asked.

Oswin had never heard the man speak before. He had the low voice of someone always in complete control of himself who demanded that others around him be the same. Like the air itself would stop to listen. "He—"

"Actually, I think I was the first," Contessa interrupted. "When I was in the river, it tried to abduct me. They scared it away with the rover when they came searching for me."

Rhona nodded in agreement.

Dr. Williams's intense blue eyes moved to Contessa. "That's when you got sick, but you didn't die like the others."

"That's right," Contessa said.

"Before the capture, did Fanar interact with this creature?"

Oswin scowled. *What is he insinuating? Does he believe that Fanar and Ciabgan are working together?*

"You think Fanar helped him?" Rhona asked. Oswin almost smiled. He could always depend on her to speak her mind. She leaned forward and raised her voice. "Why would you think that? If anything—"

Dr. Williams held up his hand. "I'm simply trying to learn about the time they spent together." He looked at the other two board members. "If it can excrete a toxic substance, perhaps it can also excrete something that subdues its victim? He may have been acting under a chemical influence."

Dr. Chernoff raised his thick eyebrows. "I had not thought of that. Such a thing may be possible. Because we're from another world, it could act on us differently than it would the native animals."

Nobody spoke for nearly twenty seconds as that sank in. Fanar acting out because he was under the influence of Ciabgan? Like a hypnosis? Oswin had his own suspicions about why Fanar seemed to favor the Nitonian. Oswin had even accused him of letting Ciabgan go. Of course, Fanar had yelled at Ciabgan and gotten the creature shot. Oswin shook his head—so much to take in. "What if—"

A soldier opening the door stopped him from speaking. "There's a fire down by the lake."

"Fire? What's over that way that can burn?" Dr. Yamamoto asked. Oswin's eyes went wide. "The water pump!"

"But it's in the water," Dr. Williams said. "How could it burn?"

"It's that infernal experimental converter," Dr. Chernoff said. Without a word, he followed the soldier outside. A rover stopped outside the entrance. Dr. Chernoff talked with the driver then motioned for the others to come. All six of them squeezed into the rover with Kellach, who was driving. Dr. Yamamoto sat in the back with Oswin and the sisters, letting the men sit in front. They were barely inside before Kellach put the vehicle into motion.

Contessa fell onto her sister. She whispered just loud enough for Oswin to hear, "Did Fanar take driving lessons from Kellach?"

Once they were away from Landing, the orange glow of the fire could be easily seen. As they got close, the bright orange color shifted to a dull orange with a slight blue hue at its base.

"I think it's going out," Rhona said. "That's good."

Kellach stopped the vehicle and got out. Not waiting for anyone, he approached the burning equipment. He got as close as the heat would allow and circled it, studying it from various angles. He pulled the burning hoses away from the pump and stomped out the flames at the ends as he growled, "Of all the times to fail."

Dr. Yamamoto walked over to Kellach. Water was spilling out of the charred ends of most of the hoses. "Let's see if we can use this water to put out the fire." She put words into action by grabbing one of the hoses and waving it toward the pump, splashing wave after wave of water toward it.

Kellach watched Dr. Yamamoto and Oswin wave the hoses so the water sloshed onto the fire. "It's not going to work," he snarled.

"Why not?" Dr. Williams demanded.

"It's worse than I thought. We need to find a way of pulling it out of the water. See the blue tint? That's hydrogen burning." Dr. Williams shrugged.

Dr. Chernoff dropped the hose he was holding. "That converter will keep splitting the lake water. The heat will only accelerate the conversion to hydrogen and oxygen."

"You know about this, then?" Kellach asked.

With a wave, Dr. Chernoff dismissed the notion. "I only know the basics. Nothing that can be of real use. How do we stop it?"

Kellach took charge. "Tie knots at the end of those hoses. I need as much water to stay in those bladders as possible." He paused and stared at the burning heap. "I don't know if I have anything that can help. The pump is on wheels; maybe I can find something in maintenance to pull it out." He stormed to the rover, his feet spitting rocks and dirt behind him.

"I can come with," Oswin volunteered and followed. "No, I got it."

"I think multiple people would help. Why not all of us?" Dr. Williams suggested.

Kellach turned on them and snapped, "No. I have some sensitive things happening in the building, so I need to keep everyone clear for a while." He could be heard muttering to himself as he entered the vehicle, "This is a disaster."

Oswin watched the retreating vehicle. *Since when is no one allowed in the maintenance building?* He thought back to Fanar's comment about locked doors. Now Kellach or one of the soldiers would bring items in or out of the building for whomever needed them. Even the rovers would be personally delivered instead of letting the user simply take one out as they had done when they'd first arrived on Niton. Rhona stood beside him with her hands on her hips, fingers facing back. Oswin waited for her to say something, but nothing came. Perhaps nothing needed to be said; her stance said it all.

Dr. Yamamoto tried splashing more water on the burning heap to no effect. After nearly fifteen minutes of toil, she gave up. "It's only getting worse."

"I don't get how it caught fire," Dr. Williams said.

Dr. Chernoff shook his head. "I've been thinking about it, and I don't know either."

"I bet it was sabotaged," Oswin said. "Fanar's out trying to find Ciabgan and it or one of its kind is still here sabotaging our stuff."

"How do we protect ourselves from something that can become invisible?" Rhona asked.

Seemingly from nowhere, Kellach drove past them into the water and maneuvered the rover parallel to the shore. With a splash, he jumped out of the vehicle. "Help me with this cable," he yelled out as he sloshed his way toward the back of the rover.

Oswin ran over and helped pull a dark cable out. It appeared to be the same kind used by the platform cranes. While he pulled the cable off, Kellach secured one end to the rover.

Kellach motioned with his arm as he talked and moved away. "Now a couple of you wrap the cable around the pump, and we'll pull it out."

Rhona and Dr. Chernoff helped Oswin pull the cable in a wide arc around the burning pump. At one point, Dr. Chernoff dunked himself into the water so he could tolerate the fire more easily. It was such a good idea, Oswin and Rhona did the same. After wrapping the cable around the pump two times, they brought the other end back to the rover and handed it to Kellach.

Kellach peered into the darkness toward Landing. "What's taking him so long?"

Oswin stared in the same direction. *Who is he waiting for?* After several seconds of looking, he saw the lights of a rover bouncing toward them. He had to get out of the way as a soldier moved the other rover into position parallel to the one Kellach had brought.

"I hope this works," Kellach said as he secured a cable to both rovers before he and the soldier slowly pulled the vehicles forward until the cable tightened then stopped.

"Why are you stopping?" Oswin yelled from his vantage point.

"I'm not!" the soldier yelled back. He tapped his panel, and the vehicle moved forward a few centimeters then suddenly lurched forward.

Oswin watched the pump, expecting to see it moving out of the water. Instead he saw the cable flying through the air. *It broke?* He ran over to it and pulled it out of the water. Rhona rushed to his side and helped. They drew out the entire cable without seeing any breaks.

Dr. Chernoff came over to look. "The heat must have caused it to break."

"No, it didn't break. See?" Oswin gestured to the cable that lay at his feet. "It must have slipped off."

With a sigh, Dr. Chernoff picked up some of the cable. "Here we go again."

Oswin took the end from Kellach, who detached it from the rovers and headed into the water. As Oswin rounded the far side of the pump, he stopped. "Maybe we need to get it underneath."

"How can we do that?" Rhona asked. "It's too hot."

Oswin didn't have a clue. How hot was the water under it? Probably too hot. As predicted by Dr. Chernoff earlier, the fire had gotten bigger. He was surprised the tires hadn't already blown out. Tires? They were big enough— maybe they could loop the cable around them. "What about the tires?"

Dr. Chernoff nodded. "That may work. If we can loop it up under and secure it against the axles, then we can pull it out."

It wasn't what Oswin had been thinking, but he agreed. They pulled the cable across and down into the water and moved back to shore so the cable came up into the wheel well and against the axles.

Kellach snatched the end from Oswin and attached it to the rovers. In seconds, he and the soldier were pulling the cable tight. This time, it held, and the pump crept toward land. The intensity of the fire diminished as the bulk of the pump came out of the water. When it reached land, Kellach and the soldier took the cannon-like fire extinguishers to the remaining flames, causing the area to grow dark. Kellach glared at the pump, yelled an expletive, and kicked one of its tires. He glanced at Oswin. "He'll take you guys back. I gotta see how much water we lost."

Oswin looked at the ruined hoses. First it was parts for the houses, then it was sensitive equipment, then the water bladders, and now this. And what of Dr. Williams's suggestion of Ciabgan's influence on Fanar? Could he be in danger? He had a gnawing feeling Fanar was in great danger.

# CHAPTER TWENTY-FIVE

## Rescue Mission

The soldier dropped them all off at the Kings' house. Contessa addressed

Oswin. "Rhona and I have the day off. Are you okay?"

"I'm worried about Fanar. What if Dr. Williams is right and he's been acting strange because of Ciabgan?"

"Have you met Fanar?" Rhona asked. "He's been strange from the day we met him."

Contessa coughed into her hand in a futile attempt to hide a smile. "He lost his mom last year; it's been rough on him," Oswin said.

"Really?" Contessa asked. "I had no idea. I thought maybe his mom and dad were divorced. And to lose his dad…"

Oswin nodded. "He doesn't like to talk about it. I'm afraid for him. If Ciabgan can influence him, who knows what he'll do? Are they recruiting him to do more against us?"

"They?" Rhona asked.

"There has to be more of them out there. I think the first one we saw, the one that became invisible as we watched, was not Ciabgan. It was too far away from where we found Ciabgan."

Rhona raised her hands questioningly. "So what can we do? Chase Fanar down and convince him to stay with us until they open the portal?"

Oswin sighed. "I don't know. He's convinced himself that Ciabgan knows where his dad is. He could be halfway across the continent by now." Contessa grabbed his hand and caught his eyes with hers. Her brown eyes stared at him in earnest. "Then we'll go all the way across. What do we do?"

Thankful for their support, Oswin gave her question some thought. If they were to find Fanar, they would need a rover. If there were no problems, the battery could last for four days of constant traveling. Half the day was gone already, but they needed to leave as quickly as they could. "We need to get a rover. One that's completely charged, and we'll need to ask what direction he went."

Contessa squeezed his hand and let go. "You leave the rover up to us. You find out where he went."

He nodded and walked toward food processing. They had been the last to see Fanar and could tell him what direction Fanar had gone. The door to the building was locked. He had really hoped a small group of people coming to an unpopulated world would be able to keep their doors unlocked. The Nitonians had certainly squelched that dream. Now which house was Dr. Yamamoto in? He crossed the street and knocked on the door of G-1.

Dr. Yamamoto opened the door. "Oswin, what can I do for you?"

"Hi, sorry to bother you. When you saw Fanar leaving last night, what direction did he go?"

She smiled. "Going after him?" He nodded. "He went up the road in the same direction as we did the harvesting. Rhona can show you."

"Thanks." He turned to leave. "Oswin?"

He spun around.

"I hope you find him. At least he has the rover for shelter. I don't believe that creature has any control over him. He's grieving. I'm sure he'll be okay."

He nodded and headed back toward his house. She might give herself comfort in that belief, but he wasn't so sure. From the time Fanar had started to search for his dad, he had been obsessed with the Nitonian and finding his dad. There was no way he would come home on his own without his dad.

Oswin went into his house and packed some food. He had just finished when Rhona and Contessa let themselves in. He peeked out from the kitchen. "Need me to help with the rover?"

Rhona grinned. "Nope."

His eyebrows shot up. "That was fast. Hold someone up?"

"I would have, but Contessa flirted with one of the soldiers and convinced him to stay here."

"Well, that's one way. Fill these with water." He pointed to some containers he had set on the countertop. "Dr. Yamamoto said Fanar went in the direction you were gathering food. You know the way?"

Rhona paused. "Um, sure!"

"Good." He waited until all the containers were full and put most into his pack. "If we hurry, we can search a large area before we get too tired." After they organized the items into the three packs he set out, they went outside and waited. After thirty minutes, he was wondering if the soldier had forgotten when a rover's headlights made their way down the road and stopped. The soldier climbed out of the rover, gave a smile to Contessa, and walked away.

"Service with a smile," Rhona said.

The back of the rover had additional equipment already loaded. A nice addition. That soldier must have taken a liking to Contessa. Oswin got into the driver's seat, and Rhona and Contessa sat next to him. He tapped the console. The charge was only at ninety percent, but that would probably be enough. He looked at the map but only saw a couple of the food sources. "Is it one of these?"

"I don't think so. We sort of went straight up and over a little that way." She pointed on the map next to some hills where no indicator had been placed. "That's where I saw some orange berries."

He nodded and started the rover moving. It didn't seem too far away. As they rode, his mind wandered to Ciabgan and all that had happened since they'd arrived on Niton. Randy might have been right when he said force was needed. That electric net had done the job, and it hadn't killed the creature. When they opened the portal, maybe more soldiers could be brought back. Dr. Monier was wrong about not needing them. The colony needed more. They needed soldiers at the different worksites and some to patrol around Landing. It was the only way of keeping safe against the constant damage the Nitonians were causing.

Rhona studied the landscape outside the window. "This doesn't look right." "It's dark, and it's getting late," Oswin said. "Hard to tell what it's like. Tomorrow will be a bright day, and we'll see better then."

Rhona yelled out, "I get back seat!" She scrambled over the top of the seat to get there before her sister had a chance to respond. She stretched herself across the entire seat and grinned. "This will work just fine."

Contessa snorted and lay back against the front seat, trying to get comfortable.

Oswin leaned against the window. Rhona definitely had the best spot. It wasn't long before he slept.

When the sun woke him, he found Contessa had shifted to lie across most of the seat, her head next to him. Rhona snored in the back. He got out of the vehicle and explored. The landscape was dark, pitted, and void of vegetation. *This can't be the right spot.* A mangled tree with half its foliage fighting to remain relevant stood alone about fifty meters away. He walked to it. The trunk was split and scarred. He had no way of knowing how old the tree or the damage was. Maybe Dr. Yamamoto would know.

"Oswin!" Rhona called.

He pivoted on his heel. In the opposite direction of the rover, Rhona and Contessa stood near a grey boulder. He jogged toward them and slowed as he approached. The boulder resembled a large inflated Nitonian. If it were brown, he would have been running the other way. He came to a stop next to Rhona. It looked exactly like a Nitonian!.

"Is it—?" he asked.

Rhona rapped her knuckles on it. "Hard. I saw this before and thought it was going to attack me."

He could hardly believe it. It even had the six bulges on the sides where the outer covering would wrap the legs and a bigger bulge in the front for the hood that would cover the head. He tapped the side with his fist. It felt solid. He didn't see any others. "Maybe it's a statue?" The words came out and sounded silly to him.

"Why would there be a statue out here?" Rhona demanded.

He shook his head. "I know, I just… I can't think of what it might be." "Maybe it's a tombstone," Contessa said.

"Weird. So where's all this food you were gathering?"

Rhona looked around. She stood next to the statue and turned so it would be on her right and scowled. "I don't understand. This was here, and there was a hill…" She swallowed. "How did we get on this side, anyway? We were supposed to stop before the hills." In a huff, she stomped her way to the rover and got in.

He chuckled to himself and followed. The sun had already set when they'd arrived the previous evening, so they'd probably ended a little off course. "Why don't we just start searching for Fanar now?"

"We should start where I was gathering food. That Ciabgan creature had to eat, and Fanar followed it," Rhona said.

He couldn't argue that logic, so he pulled up the map display. "Okay, this is where we are now. Where should we go?"

She stared at the screen. "Can't we just drive around until we find it?" "You don't remember, do you?" Contessa asked.

"Of course I do!" Rhona snapped. "I just wasn't watching the map as we rode in."

"It's okay; we can drive around. It's no problem," Oswin said. The last thing he needed was to listen to them fight. He manually weaved the rover between the hills for the next half hour. The landscape didn't change—a few living plants with a ground that appeared to have been scorched. After another ten minutes, he stopped. "Rhona—"

"Yes, I know. Let me think. Pull up that map again?" She watched as he showed the map.

Contessa leaned over and pointed to another set of hills to the east of their location. "Maybe you meant these?"

Rhona slapped her hand away. "I know what I meant. Let's drive around some more. I'm sure we're close."

It was pointless, but he drove around some more of the hills in an attempt to find something, anything, that resembled plants that could be harvested for food. As they roamed in the rover, Oswin's eyes scanned the horizon for any sign of Fanar or Ciabgan.

"Look out!" Contessa yelled.

It was too late. They lurched forward against their seatbelts as the front tire dropped into a crater and the rover came to an instant stop. He undid the clasp of his safety strap and went around to assess the damage. He cursed himself for not paying attention. Like when he was in training, he had forgotten to direct the vehicle. The rover's body was damaged, but nothing else seemed to be wrong, so he got back inside. With a squeal protest, the rover backed out and came to a stop. He pulled up the map and tapped the area Contessa had pointed out earlier.

"What are you doing?" Rhona asked.

"We've been all over this area at least three times," he said. "We need to search elsewhere."

She sat back and stared ahead as the rover creaked to its new destination. The rover bounced as they drove through shallow craters and hit small rocks. When they arrived at the new destination, she leaned forward and stared for a minute before sitting back. "You know, these hills all look the same. It isn't my fault I got the wrong ones."

"So it looks familiar?" Oswin asked. He was opening his mouth to say something else when he winced in pain as a piercing screech screamed from the front of the rover. Then, with a pop, the front lurched up before hitting the ground with a crunch that brought the rover to an immediate stop.

Oswin Carter's log entry. Rhona, Contessa, and I decided to rescue Fanar from Ciabgan. That was two days ago. The front axle of the rover broke, and we have no way to repair it. (System mark: Day 60)

"The portal opens this afternoon," Contessa said.

"At least we won't run out of food." Rhona took another bite of the blue leafy plant they had picked. It was something previously defined as edible that grew in abundance.

Oswin absently chewed as he stared into the hills. "Rhona, you said you saw a boulder and you thought it was a Nitonian?"

"Yeah, right around this hill," she said.

"Think you can show us?" "Um, sure."

He held her gaze. "Really?" He didn't want to spend an hour walking for nothing.

"It was dark before! But now it's light, so it won't be a problem." She got up. "Follow me." With confidence in her step, she led them from the area where plants grew abundantly around a hill, onto torched ground, and past another hill until they could see the boulder ahead. Like the boulder they'd seen earlier, it could easily be mistaken for another Nitonian if it weren't grey.

Oswin rapped the top with his knuckles. "There has to be a reason for this. I just can't think of what it might be." He stepped back and scrutinized the boulder. The hood was suspended several centimeters above ground. He got down and peered up into the hood from underneath. His eyes went wide, and he fell back.

Rhona gasped at his action. "What? What is it?" She looked under and gasped again. "Is that—"

"This isn't a statue. It's a dead Nitonian." He stood and more closely

examined the surface. "It's not petrified, but something similar." Contessa stood back. "You sure it's dead?"

Oswin tapped on various points of the shell. "Sure seems that way." "Shhhh," Contessa said. "I heard something."

He strained his ears but couldn't make out anything beyond his breathing. "What did you hear?" "I don't know."

He listened for another ten seconds before giving up. "I don't hear anything. Maybe—" He heard a faint voice. "Maybe we should get back to the rover." The others didn't need encouragement. They followed Oswin back toward the rover. When they rounded the last hill, Oswin could see three people standing near the rover.

Dr. Yamamoto visibly sighed. "There you are. We've been calling for you. You've been gone a long time, so we decided to come for you."

"Fanar's back?" Contessa asked.

"I'm sorry, Contessa, none of the teams found him." She pointed to the broken rover. "What happened?"

Oswin could feel his face flush. "I hit a hole."

# CHAPTER TWENTY-SIX

## Encounter

*This is Fanar. (System mark: Day 58)*

*This is Fanar. I've stumbled onto a large group of Nitonians, and Oswin was right. I'm afraid they're going to kill me. (System mark: Day 58)*

anar slapped the uplink and lay against the wall. The room was a large round chamber with a ceiling shaped in a perfect half sphere. Several globes dotted the wall halfway down from the top, giving off a soft white glow. The room spun, and the next moment he realized he was lying on the floor. *Probably for the best.* Whatever they had done to him made it difficult to focus. *How long have I been here?* He didn't have the energy to check the uplink. He wasn't even sure why he had done the log entry; nobody would hear it.

He closed his eyes. After several minutes, he heard the soft thumping of padded feet walking near and past him. When he opened his eyes, he saw two small heads peering at him from the chamber's arched entrance. The Nitonians were much smaller than Ciabgan. Curious about the man animal being held captive, no doubt. Or were they comparing him to his dad? Surely his father was being held somewhere. If he could escape and investigate, then maybe he could find his dad and they could get back to Landing.

A member of the community stopped and communicated with the small ones. They clicked and whistled back and left. *That's right, stay away from the dangerous Earthling. You know, the one who just wants to find his dad and keep you from destroying everything.* Fanar closed his eyes. Each heartbeat pounded through his skull. Who was he kidding? He could barely stand. There was no way he could find his dad and escape, let alone prevent them from destroying all of Landing. More padded feet. It sounded like there were several of them. He opened his eyes.

Pouring through the archway were dozens of Nitonians. They were much larger than Ciabgan, and while most were brown colored, some had a gold tint. *Females?* They moved to the sides and made room so even more could come in. A public execution, then. He pushed himself into a sitting position. He was the intruder of these lands, but it didn't seem fair. The influx of witnesses came to an end. A hundred or more stood at the sides, making a clear path between him and the entrance. *I could try to make a run for it.* The thought was squashed as another entered and stood in the center of the archway, holding a large metallic object. *The missing drone!*

Their expressions were indiscernible as the chamber echoed and amplified the roar from the creatures. Probably cheering for their leader, encouraging him to proceed with the rightful destruction of the one who had brought this foreign object onto their land before invading it. The leader moved closer and placed the drone about a meter in front of Fanar.

The Nitonian glared at the others, and all grew quiet. He nodded his head to the drone and started a monologue that only his kind would understand with pauses and some quick, urgent gestures with a tentacle. He talked to those around him, gesturing at Fanar from time to time. The shape of the room made it easy for everyone to hear him. He nodded at

Fanar and then back at the drone. After some more clicks and whistles, he came to a stop. All was quiet.

The leader strode toward the doorway and stopped, leaving Fanar dumbfounded. *Is he done making his case? Am I now in the hands of an angry mob?* The leader moved, and Ciabgan came in. He approached Fanar with the leader following. Fanar identified Ciabgan by his evident gunshot wounds. He was glad to see Ciabgan's color was now brown like the others. Fanar and those in Landing had hunted him down, imprisoned him, and then shot him. Reason enough for Ciabgan to be the executioner.

Ciabgan walked around the drone. "Shanar. You come." He lowered himself down until his legs were hidden by his protective shell. He watched Fanar and waited.

Fanar was confused. *Am I supposed to respond?* "Yes, I came to find my dad."

"Landing."

So much had happened at Landing; a lot of it he regretted. "I am so sorry you got shot. I never wanted that. You gotta believe me. I just wanted you to give my dad back." *Does any of this make sense? How much English did he pick up while visiting Landing?*

Ciabgan whistled from a high to low pitch then clicked. He closed his eyes for a couple of seconds then opened them. "Questions." "I have lots of questions."

Ciabgan waved his head back and forth like a snake under the influence of a snake charmer then leaned forward. "Questions."

Now Fanar understood. They had questions for him, and Ciabgan was to be the interpreter. He swallowed. Coming onto Niton—that, he could defend, but there was no way he could defend the shooting. He nodded. "Okay."

Ciabgan pointed at the drone. "How?"

Fanar scowled. *How? How what? How was it made? How was it used?*

"How come to Landing?"

They wanted to know that? Not why their lands were being invaded or when the colonists would be leaving? "I don't know much, but I can say we created an artificial wormhole and came through that." Fanar waited, but only got a stare. "A hole through the fabric of space."

Ciabgan turned toward their leader and said something in his language. Immediately chatter filled the chamber. He turned back to Fanar, and all grew quiet again. "Why?"

This question, he expected. He swallowed. "Resources. Fresh water and land for food. We did some things wrong, and our water got polluted with microscopic pieces of plastic and chemicals, and our technology used up all our rare minerals. We got some technology for travel and found a way to adapt it to come here and other places." *That's a lot of information. Did he understand any of it? What's going to happen now?* Twenty agonizing seconds passed.

"Drink water."

"Yeah, you could say that."

Again Ciabgan interpreted for the others, and the chamber erupted with noisy chatter before they all moved.

*This is it.* Fanar closed his eyes. His heart quivered. He waited, but nothing happened. He opened his eyes to see all of them, except Ciabgan and another, leaving the chamber. *Wait, I'm safe? That's it?*

The other one chatted at Ciabgan, glanced at Fanar, then pointed at Ciabgan's gunshot wounds. He paused while Ciabgan responded. The other one said something else before he glanced at Fanar again then left. Was he against sparing Fanar's life, especially in light of what had been done to Ciabgan? *Maybe I'm not safe after all.*

Ciabgan reached out with one of his tentacles and touched Fanar's head. "No see." He rose to his feet and left.

Fanar watched him leave. *What am I not seeing?* After Ciabgan left, one of the smaller Nitonians came in and set a glass of water next to him. Drink the water—that was what he'd agreed to. The glass was clearly one brought from Earth. Probably taken when the Nitonians had been sabotaging things that first night. He was thirsty, so he drank it and lay down. His head still hurt. Maybe some rest would make it better.

When he woke, he found more water and some leafy blue plants on a thin platter made of stone. The plant looked like something Dr. Yamamoto had recently introduced to Landing that he had not tried yet. Perhaps he wasn't going to die, but the interaction between Ciabgan and the other after his interrogation didn't sit well. He nibbled on the plant. It wasn't terrible, and his head felt better. There weren't any Nitonians in the chamber; maybe he could take this chance to leave before they realized their mistake in keeping him alive. He stood and sidled along the wall toward the exit.

A quick peek out revealed a tunnel to the left and right with nobody in sight. He had no idea where he was, so he went left. The tunnel was smooth and arched, much like the chamber he'd left, with globes placed on the ceiling at regular intervals to light it. A warm current of air caressed his skin. He passed a few small chambers, some of them lit like the one he had been held in and others dark. No sign of his dad. Ahead of him, a couple of large Nitonians, one brown and the other gold tinted, exited a chamber with a young one between them. Fanar froze.

The trio went ahead of Fanar in the same direction, so they didn't see him. *Are they going to see my dad? That could explain the empty passages. Everyone went to see my dad and question him.* Keeping a safe distance, he kept behind them down some adjoining tunnels and past a few more empty chambers. Some were rather large, and one had an array of bowls with liquids and crystals of different sizes and colors. It smelled like—his heart froze. It smelled like an invisible Nitonian! His feet faltered and stopped. He swallowed and held his breath. There was no sound of being followed. The Nitonians he was shadowing turned down another tunnel.

Fanar wanted to run and hide in one of the rooms, but he also didn't want to lose the ones he followed. *What should I do?* He pictured the Nitonian who'd last spoken with Ciabgan lying in wait to kill him. He let his breath out slowly and wiped his sweaty palms on his pants. He looked back to see an empty tunnel. Not hearing anything, he made up his mind. As quietly as he could, he ran to the tunnel the Nitonians had disappeared down and saw…nothing. *Where did they go?*

As he contemplated their disappearance, he heard a low hum. It came from a doorway down the same tunnel to the right. He tiptoed to the doorway and peeked through. Inside, a mass of Nitonians stood staring forward with an empty aisle in the middle leading up to where the leader stood. Was that a statue next to the leader? It resembled a sleeping Nitonian, but all grey. Just inside the doorway stood the ones he'd tailed. The young one walked forward a few meters then ran back. The female leaned in and whispered something to the young one while caressing the face. The young one turned, walked down the aisle, joined the leader, and faced the throng. The leader gave a pat on the young one's head.

The hum quieted, and the young one began to whistle a melody. The whistle was a sweet counterpoint to the hum that now changed and grew in intensity. Fanar took in a shuddering breath as a wave of emotion swept over him. Watching, he realized that this was a funeral and that the Nitonian-shaped object wasn't a statue, but the one who had died. A figure came into view and shuffled toward the dead Nitonian. He was nearly as large as the leader, but his shell color had hints of grey, and his head was lined with deep wrinkles. He appeared very old to Fanar—perhaps the elder. With deliberate movements, he placed crystals onto the carcass, then some liquids. The crystals began to glow. He reached down and struggled to lift something. The leader assisted, and they brought up a highly reflective domed metal sheet. Together, they placed the dome over the dead Nitonian. One of the Nitonians near the back turned his head. Fanar jerked back, his heart pounding. Wiping unexpected tears from his face, he tiptoed away and went down an adjoining tunnel. *If they're all here, then now is the best time to find Dad and escape.* With no sign of being followed, he quickly scanned each room as he ran through the tunnel. The darkened rooms took a little longer to scan, but light from the tunnel kept it from being an impossible task. Some of them had bowls of food and water, while others had crystals and foul-smelling liquids. He came to an intersection and turned right. *Is this the way to where I was kept?* Except for the size, the rooms all looked alike. The tunnel ended.

Fighting down panic, he retraced his footsteps and went in the opposite direction at the intersection. A few large rooms here with two pools of water in each. *Public baths?* This tunnel too came to an end. He raced to the previous intersection and ran past it. This was familiar. *How big is this place?*

He had easily scanned over a hundred rooms, and his dad remained undiscovered. He had to be close. The chamber to the right was large and mirrored the one he'd been in earlier, but then, so had many others. He continued past and came to another intersection and turned left. Three more chambers and a dead end. He ran back to check the other direction and came into a large chamber filled with food and water. Shelves of stone that seemed to grow from the walls held enough food to feed Landing for a week. On the other side, another doorway. He ran through to encounter a maze of tunnels. He stopped. How far did he risk going? The funeral could only last so long.

Fanar glanced behind himself. Not seeing or hearing anything, he slunk into the new section. After what seemed an eternity of finding rooms resembling others he'd passed earlier, he ended back at the food storage chamber. Only one direction remained, and that would lead him past the funeral gathering.

He still hadn't seen the large crystal cave where he had first encountered the group of Nitonians. He realized he must have missed parts of the tunnel system. His breath was loud in his own ears as he came near the gathering. He stopped running and controlled his breathing. Turning his ear toward the opening, he couldn't hear anything from within. A quick peek revealed that they were all still present, staring ahead at the leader, who held his arms high with his own eyes closed. The dead Nitonian was nowhere to be seen. Fanar didn't have time to ponder that last part. He held his breath and ran across to the other side then sprinted down the tunnel. The next chamber was different. It was a half dome like the others with liquids and strange- looking objects on shelves, but a tower of stone with crystals of various colors stood in the middle. Tubules of what seemed like glass ran in and out of the tower, pulsing light at random intervals. Blue and yellow crystals, however, all pulsed in a hypnotic rhythm like a heartbeat. The blue then the yellow followed by a pause, then the blue and the yellow again followed by another pause. Time seemed to stand still as he gawked at the apparatus.

"Shanar."

Fanar jumped and turned to see Ciabgan.

# CHAPTER TWENTY-SEVEN

## The Light

"Ciabgan!" Fanar exclaimed. His mind raced, and he blurted out the first thought he could grab. "I thought maybe I could go home, but I think I'm lost."

"You no see. You see, go Landing," Ciabgan replied.

*What? What am I not seeing?* He followed Ciabgan down the tunnel, peering into the different rooms again. The air had a different quality as they passed the funeral chamber. It was pungent with the odor of ozone and a hint of charred flesh. Had he witnessed them cremating the dead? The tunnels were vibrant with activity, and his passing drew many stares. He swallowed. *Is that good or bad?*

They didn't go into the room he'd woken in. Instead, they went past it and entered a small chamber where the leader waited with the elder. They were talking to each other and stopped when Fanar entered. The leader gazed at him, and Fanar realized this might have been the one who had hurt him when he'd entered that crystal cave. He steeled himself for the pain.

The leader looked at the elder, said something, and stepped back to allow the elder to come forward. The elder touched Fanar's head with one of his tentacles and chattered in his own language. The tentacle caressed Fanar's face. When Fanar stiffened at the touch, he dropped it and motioned for Ciabgan to come forward and talked to him for nearly a minute before turning back to Fanar.

"Now you see?" Ciabgan said. It was almost a question laced with doubt. Fanar shook his head. "Sorry, I don't see. What should I be seeing?"

Ciabgan stared into Fanar's eyes while the elder laid tentacles onto each of them.

A focused point of pain formed inside Fanar's forehead. "You're hurting me. It hurts!"

Ciabgan said something to the elder, who chatted back. Fanar was about to pull back when all went white.

"This is where you're safe, that's why," said the leader in a stern tone. But he was more than a leader; he was the adopted father.

Ciabgan sighed. Maybe the others were content to dwell in these caves for the rest of their lives, but he needed to get out. He planned to wait seven sun cycles before attempting his escape. Their kind was patient, and he knew if he went sooner, his absence would be noticed. It would be the perfect time, when the Entrusted went out to gather body protection and food that could not grow inside the caves.

He had to wait another four cycles after the seven had already passed before several of them went out. After they left, he followed at a safe distance. Once outside, he went across the water. The others would be moving in the opposite direction. After he crossed the water and got away from the trees, he ran. He would have run sooner if he could have, but his kind wasn't made for seeing far. Not with their eyes, anyway. He relished the refreshing feel of the air rushing against his face. It had been so long since he'd sped in the open space like this.

He slowed his pace when he reached the northern edge of the lake, breathing hard. He dropped his head to the water and drank. Now he needed to rest. He'd enjoyed the run, but it had made him tired. He drew air up into his outer self, pulled his tentacles and head inside the protective hood, and settled down to sleep.

Ciabgan didn't know how long had passed when he woke. He would have to return soon, or he would be missed. He looked at the lake then up past it. The sun was high—maybe he could go just a little farther first. He didn't run but moved quickly. How far could he get before he had to return? This time away from everyone was exciting! Besides the low ground covering, no vegetation grew here. No bushes or trees nearby. While he moved, the air itself hit him. Like he had jumped into the water and it slammed against his body, but it was the air. Instinctively, he inflated and pulled his head up near his body. A shimmering greyness appeared ahead.

Afraid to move, he watched and waited.

Nothing changed for several minutes, and then suddenly something came out of the grey. It floated in the air without anything holding it up. It flew past him toward the lake, beyond his range of vision. He waited to see what would happen next. After nearly an hour, the greyness disappeared. Never in his life had he heard of such a thing. He waited a few more minutes before deflating and walking to the area where the greyness had appeared. He couldn't find any marks on the ground or any other indication it had been there. *What came out? How did it stay in the air like that?*

He trotted toward the lake, following the path of the object. A square shape, about the same width as himself with a low profile, it silently sat on the ground away from the lake. After watching for a minute, Ciabgan slowly approached and tapped it with his tentacle, ready to make a quick retreat if necessary. Nothing happened. *The others need to know about this.* He would most likely get in trouble, but it had to be done. He ran back to the caves and quickly ran into his father.

"Why do you defy me?" his father asked. "I took you into my care, and you defy me."

Ciabgan hung his head in shame, knowing his father was right. "Will you keep the light from me?" If his willfulness kept him from receiving the light, he didn't know what he would do.

"It is not my place to keep the light from you. You were chosen by Aiyam to receive it and become one of the Entrusted. But you are no longer a hatchling and must be responsible." He paused. "I suppose it's partly my fault. Time has passed so quickly I failed to see you can care for yourself." He put a tentacle on Ciabgan's head. "You are safe, and this cycle, you will receive the light as ordained."

"Father, I saw something. Something you must see." He stared into the leader's eyes and projected into his mind what he had seen by the lake.

The leader stood for several seconds, taking in what he saw. "After all these years, it has come to pass. Truly the timing was perfect for you to witness this. May I share this with the other Entrusted?"

"It is now yours to share." He watched his adopted father leave and wondered what had come to pass.

It had been a while since he'd replenished his body protection, so he went to one of the rooms with pools of water. When he was a hatchling, he would roll himself on the plant that held the protection. Since they had taken refuge in the caves, the plant had to be brought in and placed in the pools of water.

Before immersing himself, he checked for any evidence of open sores. The protection was needed for removing the mineral from his skin, but if the protection got into his blood, he could die. Not seeing any wounds, he immersed himself into the first pool. A modest current of fresh water pressed against him, removing any debris that might be stuck to his outer self. After turning and exposing his entire self to the clean water, he moved to the next pool. Here, he took deliberate steps onto the plant that had sunk to the bottom to release the protection. He didn't want to go grey before his time. He sat there pondering what he had witnessed, and his father came into the room.

"Ciabgan, it is time for you to receive the light." He seemed particularly excited. "When the ceremony is complete and you have recovered, find me. We have much to do." He led Ciabgan to another room, where several others stood. Crystals and liquids and tools lay ready.

"Will it hurt?" Ciabgan asked.

"I will not lie—there is considerable pain. But for the chosen few, it is our heritage and our privilege to be bearers of the light. After you master its use, the pain will no longer exist."

Ciabgan accepted the answer and stood ready. They proceeded to cleanse him of his protection. He wondered why he had bothered dipping himself in the water. The process took hours to perform and was, as told, painful. The light was a secret passed from one generation to the next. One day, he would learn the intricacies of the procedure. For now, he received the gift. The responsibility.

He was given instruction on how to use the light to make himself unseen to others. As an Entrusted, he could recognize others who bore the light and see them even if they had hidden themselves. It was a great trust given to a select few and should not be misused. Ciabgan knew from a young age this day would come. He had been told over and over how it was an honor and privilege that should not be misused. He understood the trust and value of being chosen by Aiyam. When the ceremony was complete, they again pressed upon him the burden and care of this great gift. He acknowledged what they said and left.

Everyone appeared the same as they had before the procedure. He knew they were Entrusted and had the light, because only they could pass it on. Reaping the benefits of becoming an Entrusted would take time and practice. He rested for a few hours then went to find his father.

"How do you feel?" his father asked. "Sore."

"When I received the light, I was sore for a couple of cycles. Many sun cycles passed before I could hide myself."

"How long before you could see it in others?"

"Once the pain passes, you will start to see. Then we will practice because we believe the promise has come."

"What I saw in the field?"

"A prophecy that says, 'From a thirsty land, he will tunnel through the air and restore you. He will be recognized as a bearer of the light.'"

"That's what I saw! A tunnel through the air." Ciabgan considered himself a part of something really big. He had heard the prophecy before but hadn't thought about it. "What happens now?"

"Aiyam has chosen you to bear the vision to us. The timing can't be a coincidence. You must continue what you started. Watch and learn. When the time is right, we will be restored. For now, you need to rest and heal. Then you will learn to use the light."

*Rest?* Ciabgan could hardly rest when so much was happening. The next two sun cycles were agonizingly long. Everyone in the community treated him differently now. They were more reverent toward him but still encouraged him to rest. After the time of healing had passed, they allowed him to immerse himself into the protective water again. The incisions had healed enough that he would be safe. After that, he went to find his father. When Ciabgan found him with the other Entrusted, he stopped. Unlike everyone he had passed on the way, their skin had a slight sheen he had never seen before. He was also surprised to find the object that had flown through the air tunnel in the chamber with them.

"It is unlike anything we've ever seen," his father said. "How do you feel?"

Sore. Tired because he hadn't rested as directed. Afraid he wouldn't live up to the expectations of the Entrusted. Curious about the air tunnel and what it meant. "Ready to do my part."

"Because you will bear the vision to us, it is important for you to learn to control the light. How's your vision?"

"You're glowing. I think."

"Good. Have you tried to hide yourself?"

Of course he had tried, but he remained unsuccessful. "It didn't work."
"It will take time. Many, many cycles. It's not a vision we can share, but we can guide you in your learning."

The time passed more quickly as Ciabgan listened and learned and practiced. But nothing changed except his ability to see the Entrusted more easily and the soreness had completely gone away. He asked his father about it.

"Manipulating stone and metal is easy. Becoming one with the light takes time," his father answered.

Ciabgan believed he should do better. Then, many sun cycles later, he felt a pinch of pain as a spark flashed across his skin! He spun around in place. *Am I hidden?* He ran out of his chamber and asked the first one he saw. To his dismay, he wasn't hidden. Still, the event encouraged him. What had he done differently, and could he do it again?

It took a full half cycle before he could make another spark. It actually hurt when he did it. The other Entrusted told him to expect some pain at first as his body adapted to the light. Eventually, he would be able to make himself hidden without any sparks or pain. After another two cycles, he had a breakthrough and cried out in pain as sparks leaped across his entire body. Ciabgan ran out of his chamber, nearly hitting someone. This time, he was hidden! After a few more cycles, he found it easier to hide himself, but it still took time, and it hurt.

Now that he had improved his ability to hide himself, he could go to where the air tunnel had formed and watch. Every sun cycle, he hid himself and ran to the area and watched. Nothing happened. He wondered if the air tunnel would ever appear again.

It did. Like before, the air hit him as the tunnel appeared. This time, some objects appeared on the ground, and the tunnel closed. The objects rolled around and dug holes, keeping the dirt gathered in addition to some of the nearby plants.

When the tunnel appeared again many cycles later, the objects were back in place and entered the newly formed tunnel as a few objects flew out.

Ciabgan watched objects come and go through the tunnel several times. Some appeared in the air and others on the ground. How could this be the promise to restore them? He asked his father, but he was told to wait longer. His patience was rewarded, because something new arrived.

Creatures on two legs came out of the air tunnel. Upon arrival, one of them turned and went back only to show himself again. They moved to the side and set down something that made noises. The next thing to come through made Ciabgan stumble back. It dominated the shimmering greyness and it kept coming. He sat in wonder at its size when it had finished materializing. Then more of the creatures followed. Then another one of those enormous objects came into view, followed by more of the creatures. Seeing this, he could understand how he and the others would be restored. They had objects like no other! He watched as more emerged out of the air tunnel. The very large objects would ponderously come then the creatures, moving away from the air tunnel after they arrived. He realized the noises coming from the object the first group had set down was communicating a message over and over. He would need to learn their language. The next thing he saw swept away all doubt on the fulfillment of the promise. One of the creatures that appeared—the skin was glowing!

# CHAPTER TWENTY-EIGHT

## The Promise

Ciabgan shared his vision of Fanar coming through the portal along with the equipment to the other Entrusted.

"After all this time, he has come," said the elder. "I was afraid I would pass from this life before the redeemer came."

"I don't understand," said Gocian, one of the Entrusted. "Why didn't he see you?"

This question sparked a discussion of whether the visitor held the light or not. Gocian raised his voice at the leader and elder, visibly agitated with them. Ciabgan listened for a while, but he had already made up his mind. He would follow this creature, learn the language, and, at the right time, approach him.

These new creatures didn't have crystals or anything at all like his kind, the Trociabk, had, but these that came from the other place made their own illumination. For several cycles, he watched them assemble structures. He wasn't sure if they preferred daylight or dark because they seemed to sleep and work during both parts of the sun cycle. To accommodate, he would come sometimes during light and sometimes during dark. He learned the redeemer was named Fanar. He tried to say the name, but could never get it right.

On one evening, Ciabgan settled down by the lake. He was tired and considered going home to sleep. Trying to find a rhythm of when to watch them proved difficult. As he watched, Fanar came toward the lake with another of the creatures he had learned was named Oswin.

"How deep do you think it is?" Oswin asked. He grabbed a pebble and tossed it into the lake.

"About fifty meters. That's what one of the drones recorded. You know, after the first one got lost. Wouldn't it be fun if we could find it when we go exploring?"

Ciabgan was afraid they might notice him—they were so close. He listened, making out bits of what they discussed. They sauntered away so Ciabgan stood and moved toward them. Maybe he could follow them back and listen as they talked. Fanar stopped and stared in his direction. Had he been discovered? After several seconds, they continued toward Landing. Maybe he shouldn't get so close.

Several cycles later, after he left the cave, he heard something unusual. He was told to always hide himself before leaving the caves to keep those who had made them fugitives from finding them. He had just reached the area where he would normally cross the river when he heard a whine. A different sound than he had heard before. Could this be the enemy? He had been just a little more than a hatchling when they had been forced into hiding, and the visions had never been shared with him, so he couldn't be sure.

While walking away from the riverhead, he saw Fanar and his friends inside something rolling to a stop by the river. They got out, talking to each other. Rhona took off her foot coverings and got into the water. Oswin and Fanar quickly followed. Another stood by the thing they'd come in, watching. Even with some coaxing by the others, Contessa didn't get in the water and stepped away. Curious, Ciabgan followed her progress on the other side. She stopped to look into the water and fell in! She was clearly in distress, so he ran ahead and got into the water. He pulled water into his outer self to hold himself in place and made himself visible.

Contessa hit him and glanced off toward the land. He could help her get there. He pushed the water out and grabbed her with his tentacles. She opened her eyes briefly and went limp. *Oh, please don't be dead.* He pulled her out. He found the plant that housed his skin protectant covering the forest floor and put her on top of it. He watched her for any sign of life. There, she breathed! This was good. As he watched, he could hear the whine of that vehicle nearby. He didn't have time to hide himself; it still took several seconds to focus and make it happen, so he inflated himself and hid his head and tentacles in the hood. The redeemer arrived—he would make sure she was safe.

"It's good you didn't reveal yourself," the leader said to Ciabgan. "It's not time yet."

Ciabgan joined his father in the pool of water, stepping on the plant lying on the bottom. "When is the right time?"

"Aiyam will let us know. Perhaps Shanar will tell us himself in some way. You're improving with their language?"

"Some. I can understand more, but it's so difficult to make the sounds they make. Even the way I say Shanar's name isn't right."

The leader stopped walking. "It isn't?"

"No, that's just the best I can do. I've tried many times."

"They are not like anything we have seen," his father said and continued walking the pool. "The Entrusted are still in debate of his authenticity. You've watched him. What do you think?"

"I've shared the vision with all of you."

"A vision is not complete. Your emotions and thoughts at the time are not passed on. These are also important. Your witness of the arrival is no coincidence. I feel it, and the elder does too. There's something very important about the two of you together."

Ciabgan considered that. Visions were simply what was seen and heard. However, being near Fanar overwhelmed him with an odd mix of emotions. "Father, I think I can feel his emotion."

The leader's eyes went wide. "Quick, you must come with me." He left the pool and nearly ran down the tunnel.

Ciabgan caught up to him. "What's wrong—"

"Don't say a word. Not until I say it's okay." He found the elder with some others and whispered in his ear.

The elder glanced at Ciabgan then turned to the others. "I need to inspect something for our leader. Ciabgan, you can help."

Confused but remaining silent as he'd been told, Ciabgan followed the two of them out of the constructed dwelling and into the natural cave that led toward the outside. Surely the elder wasn't going outside. While he was an Entrusted, he had grown grey and could no longer hide himself. They walked to an area where Ciabgan and some of the younger ones would play when they wanted to escape the confines of their dwelling. He thought it was a secret place, but coming here now with the leader and the elder, he knew it wasn't so secret.

The leader faced the other two and nodded toward Ciabgan. "Tell him what you told me."

Ciabgan was almost afraid to say it. What was so wrong that he had to be brought to this isolated place? "I think I can feel Shanar's emotions."

The leader was about to say something when the elder stopped him and addressed Ciabgan. "Why do you say that?"

"I am happy and excited to be learning about him and his people. But when I'm by him, I have a sadness that quickly goes away with him. This cycle, after I helped Contessa to the land, I didn't have time to hide myself, so I covered. Even when I could not see him, I could tell he felt guilty and scared."

"Guilty?"

"Yes, but I don't understand why."

"I had hoped it wouldn't be." The elder turned to the leader. "It will be difficult, but there is nothing you can do."

Something was not being said. "What?" Ciabgan asked.

"I cannot keep this from you," the leader said. "You heard the prophecy."

"'From a thirsty land, he will tunnel through the air and restore you. He will be recognized as a bearer of the light,'" Ciabgan quoted.

"That's what is shared amongst our people, and it brings hope. But there's another part that's passed down only from one leader to the next." He glanced at the elder, who gave his consent, before continuing, "The next part says, 'His emotions will be felt by the one chosen to partner with him to shepherd you through the dark times.'"

"What are the 'dark times'?"

"We don't know. The full prophecy is passed down from leader to leader as ritual, but that part was never understood. Visions never contain emotion. It's… just not possible."

"This can't go beyond us," the elder said. "We don't want to cause a panic." Ciabgan could feel a rhythm to these strangers' sleep schedule, but trying to keep pace was difficult and exhausting. He lowered himself to the ground and watched as Fanar and some others gathered in the clearing. They had moved those giant things holding the lake's water. From this distance, he couldn't make out any of their faces, but the glow of Fanar's skin always made it easy to see him even in the bright sunlight. As he watched, the air tunnel formed.

Were more of Fanar's people coming? He watched the first thing holding water slowly vanish into the tunnel. They were moving the water back to their land. The prophecy said the restorer came from a thirsty land, so that made sense. Then the gathered people, including Fanar, walked in a line beside the platform into the tunnel. Ciabgan stood. *What's happening? Why is he leaving?* He got close so he could see better and hear any conversations. The platforms moved through until they had all gone. Then, a while later, a new platform emerged from the tunnel into the vacated space.

A few of the Earthlings came through beside it, then some more, followed by Dr. Monier. Ciabgan was certain he was Fanar's father. He talked to Dr. Chernoff. "I don't know what you need to do, but make sure this platform gets all the way through. Bring the motors to near overload—do whatever you can."

"What's going on?" Dr. Chernoff asked.

Dr. Monier looked around. "The power cells for the wormhole generator were damaged. We don't know how much time we have."

"Those blasted Greeners." Dr. Chernoff ran to the side of the platform, climbed up the ladder, and worked his way to a boxed-in area near the section that had just come through. After he talked to another person inside, the whining noise of the platform intensified, and it moved faster. The pungent, metallic odor of ozone permeated the air. Several minutes passed before it came completely through. Then, right behind it, Ciabgan saw the glow of Fanar. He hadn't appeared—the glow of his skin came through the greyness like a dim light through a thick fog. At that same time, the tunnel flashed, and for the first time, he felt it close.

The impact was more than any of the times it had opened. Ciabgan blinked his eyes, unable to see. After a couple minutes, his vision returned. *Where's Fanar?* He scanned the area but didn't see him. The colonists were talking so fast to each other that he had a hard time understanding. The few words he could make out were about supplies and people left behind. It didn't sound good.

Ciabgan needed to tell the others what had happened because this affected them too. Fighting fatigue, he ran toward his home. As he grew close to the cave, something on the ground got his attention. There lay Fanar and Oswin! As he approached, one of the mice saw him and scurried away. If the creature had seen him, then he must be visible. He had never been warned that he might become visible when overly tired. He closed his eyes and, after a few seconds, hid himself again. The sparks still hurt. He looked forward to the day hiding himself didn't hurt. When he opened his eyes, he saw Oswin staring at him before passing out.

For the next two days, Ciabgan watched over them. He didn't know what else to do. His father and the elder had no suggestions except to keep animals away from them. There were a couple of times he fell asleep next to them, waking with a jerk and hoping nobody had seen him. Once, he fell asleep at the opening of the cave while it was dark. He hid himself and went out to check on them. When he got to the land, the lights of a rover came on. *They must be searching for Fanar and Oswin. Maybe I can help.* As he pondered this, he saw Fanar and Oswin run up to the rover and get in.

Fanar woke with a piercing ache inside his head. The surge of each heartbeat pulsed in his forehead. And that dream, that crazy dream. Wait, that hadn't been a dream. Had it? He looked around. He was alone, possibly in the same room he last remembered. They were indistinguishable from each other; he couldn't tell. Ow, his head. Dream or not, it was killing him. He leaned against the wall and worked himself to a standing position. Clinging to the wall, he walked to the opening and peered out. He stumbled back as Ciabgan came into the room.

"Shanar. Now you see."

"I need to go. Can I go?" All he could see was that they were killing him.

"Please, can I go?"

Ciabgan moved his head near to Fanar then back. "I show you."

Fanar felt an overwhelming confusion as he followed Ciabgan. He pushed it aside and stumbled into the tunnel. The fogginess slowly cleared, and he was able to gain some sureness in his footing. They went past the rooms he had seen before. He hoped to see his dad or maybe even his mom in one of the rooms they passed. When they came to the room with the light tower, where Ciabgan had found him, he realized the truth. Neither of his parents was there. They passed a few more rooms before coming to the large crystal chamber where he had stumbled into the Trociabk. *How do I know what they call themselves?* Ciabgan led him out a different way that didn't require him leaping off a ledge, for which he was thankful.

When they got to the tunnel that led outside, Fanar sprinted as hard as he could. He ran down the tunnel and out of the opening, almost slipping off the wet edge down to the rocks below. Recovering, he ran down the ledge to the rover and jumped inside. Panting, he leaned his head against the steering wheel as tears poured down his cheeks. His shoulders shook, and soon great sobs wracked his body. His parents were truly dead. Lying out there. Somewhere. His whole body shook for nearly a minute before he gasped for breath and pounded the steering wheel.

"This stupid planet!" he yelled, spittle coming out of his mouth. "This stupid, stupid planet!" He pounded the wheel with each word. He kept pounding until a sharp pain in his hand caused him to stop. He continued to yell his frustration at the planet and how it had caused the death of his mom and dad.

# CHAPTER TWENTY-NINE

## Traitor

**F**anar wiped the tears from his face. A quick look at his uplink revealed it to be the day for the portal to be opened. He had just enough time to get back to Landing, return the rover to Kellach, and go back to Earth. He needed to get away from this place. The rover sped across the landscape toward Landing as flashes of what Ciabgan had shared sped through his mind. He shook his head and pushed the random images aside. It didn't matter. Soon he would be on Earth and he could put this horrible planet and this experience behind him.

He and his dad might have grown apart, but knowing he could never see or listen to his dad was more impactful than he'd thought it would be. Fanar remembered the time a new pulsar had been discovered and he and his mom had gotten to hear about it.

He leaned his head back and stared at the ceiling while the rover made its way to Landing, tears welling up in his eyes. He should have spent more time with his parents.

When the rover stopped, he found himself beside another rover in front of the bay doors of the maintenance building. With a sigh, he opened the door and immediately heard his name being called.

"There you are! Where have you been?" Oswin asked as he came around the back of the rover. Rhona, Contessa, and Dr. Yamamoto were with him.

Contessa ran forward and immobilized him with a hug. "I was so worried."

"I'm okay," Fanar replied. He lifted what little of his forearm he could and patted her back. "Can I have my arms back?"

"Oh, sorry," Contessa said. Before she could move away, he found himself in Rhona's and Oswin's arms. "We're glad you're okay," Rhona said. "Where were you?" Oswin asked.

"If you guys will let go of me, I can tell you," Fanar replied. He shook his arms when they were free to move. "I went to the cave to find Ciabgan and my dad."

"You were gone for two days!" Rhona exclaimed. "We went looking for you."

"Didn't find your dad?" Oswin asked, his voice low.

Fanar shook his head. "No. You guys were right. I think now I'm just going to go home. I thought the portal would be open when I got here." He gestured toward where the portal normally opened. People stood near the Oracle building, waiting for it to appear.

"You're going back to Earth?" Oswin asked. "Where will you stay?"

"I have an uncle I can stay with. It was bad enough coming without my mom, but now? I have nothing." Fanar shrugged and got into the rover. He really didn't want to talk about it. *Can't they just leave me alone?* "I'm going to pack my trunk."

"I'm glad you're safe, Fanar. I'm sorry about your father. Do you need anything before returning to Earth?" Dr. Yamamoto asked.

"No, I just want to pack and leave." She nodded, wished him well, and left.

"Mind if we come along?" Oswin asked.

Fanar just wanted to be alone, but he couldn't deny his friend. "That's fine." After the others got into the rover, he talked while maneuvering the vehicle toward his house. "They live in the caves."

Contessa's eyes went wide. "How many were there?" "Maybe a hundred."

"How did you escape?" Oswin asked.

"They, um, wanted me to see... something, and then they let me go," Fanar replied. Then he blurted, "They found the lost drone." He continued to tell about the horde that had surrounded him as their leader brought in the drone and how he'd thought for sure they were going to kill him.

"So they showed you the drone and let you go?" Rhona huffed. "How did that take two days?"

Fanar sighed. How much should he say? He was still sorting it out himself. "Well, they didn't show it to me at first. I had broken my hand light, and I was stumbling forward in the dark when I came across a cave of glowing crystals."

"Cool," Rhona said.

"Then one of the Trociabk—" "The what?" Oswin asked.

*There's that word again.* "The Nitonians. I think their leader, did

something that about killed me. It was over a day before I woke up." "What did he do?" Contessa asked.

"I'm not sure? He sort of stared into my eyes, and I thought my head was going to explode. Next thing I knew, I woke up in a different part of the caves. More like a cut-out room. All of them were like that. That's when they all came in and the leader put the drone in front of me and talked to them about me."

"Like Ciabgan? They can all talk?" Oswin asked.

"Not English. They have their own language I didn't understand. I wanted to get out of there, but I was too weak. When I had a chance, I tried to leave, but then Ciabgan —"

"So Ciabgan was there," Rhona said.

"Yeah, he was there. He told me I couldn't see. Said that once I could see, then I could leave."

"'Couldn't see'? What does that mean?" Oswin asked.

"I— I'm not really sure." The vision-sharing experience had faded like a dream. Only bits and flashes could be recalled. "He did the same thing the leader did to me, only it didn't hurt as much. I woke up sometime later and they let me go."

No one in the rover spoke for several seconds. "It's a wonder you're still alive," Rhona said.

"They destroy our stuff then nearly kill you? We've gotta get some help from Earth to protect ourselves," Oswin said.

"We discovered something about them too," Rhona said. "When they die, they get hard and grey."

Fanar turned his head. "Where did you see that?"

"Up where I was gathering some food the other day. We thought maybe you were up that way, so we went there for you. We saw one of them. It was all blown up like when they sleep, but hard and grey."

"I was able to look up under the front and saw the head," Oswin said. Fanar realized he had been right. He had witnessed a funeral.

"I'm sorry you didn't find your dad," Contessa said as they entered the house.

"Thanks. Right now, I just need to get away from here." Fanar went into the kitchen and got a collapsible trunk from the center storage area before leading the way to the bedroom. He packed the trunk with his clothes and a few other items then stared at the picture of his mom on the wall. She would understand how he felt if she were alive. His dad had brought her necklace to Niton; somehow it seemed fitting that a permanent image of her stayed here too.

He closed the trunk lid and surveyed the others. *This is it.* While he could always visit, today somehow seemed final. Oswin had been his best friend for as long as he could remember. The one he had adventures with and confided in. And surprisingly, he'd gained two new friends in the last couple of months. Rhona with her quick comebacks and zest for life and Contessa with an always-ready helping hand. He would miss them.

"Here, I'll help you with that," Oswin offered. He grabbed the handle on one end, walked with Fanar out to the rover, hefted the trunk into the rover, and stood back.

Rhona gave Fanar a hug. "Don't be a stranger."

"I won't," he replied. He turned to give Contessa a hug, but she stepped back.

"You can give me a hug at the portal. I'm coming to see you off," she said.

"It's okay. I can do this by myself."

"Maybe, but you don't have to be by yourself. You shouldn't be by yourself. I'm coming with you to drop off the rover, and you can give me a hug at the portal."

He nodded and got into the rover while she ran around to the other side and got in. He asked Oswin and Rhona, "You guys coming too?"

Oswin took a step, and Rhona put her hand on his arm and spoke up. "No, you guys go. Take care, Fanar."

Fanar waved and engaged the motor. The rover careened onto the roadway and sped around the circle toward the maintenance building. When he arrived, the bay door was closed, but the other rover he had seen earlier was gone. He and Contessa got his trunk out and set it on the ground. "Maybe Kellach is inside," he said and led them around to the front entrance.

They entered through the door. Kellach had his back toward them, putting some data discs into a case. He jerked and turned as they entered. "Fanar! You're back."

"Yeah, I just got back." He nodded his head toward the back. "I have your rover."

"Good. Good. Me, I'm just getting ready for the portal. Have a lot of information to send back." His gaze fell on Contessa. "How are things with you?"

Contessa stepped closer to Fanar. "Okay."

"You've fully recovered from your sickness?" He didn't wait for an answer. "Good thing you were left on that ground covering; its spores are a natural antitoxin for that insect's sting." His shoulders stiffened, and he turned. "That is, I assume that's the case."

Fanar's eyes went wide. "What—?"

"Nobody told me anything about an insect," Contessa said.

"You knew what made her sick and how to cure it?" Fanar asked. His jaw dropped as he suddenly realized the appalling truth. "You got everyone sick."

Kellach grabbed the disc case off the counter while bolting around it. The door leading farther into the building slammed against the wall as he went through. Fanar took chase. He jumped over the counter and sprinted down the corridor. Kellach pushed a door open and continued past toward the back.

Fanar gained on the fleeing man. He briefly glanced into the room Kellach had opened and stopped in disbelief. Lying inside was the bound and bloodied figure of his father. "Dad?" He walked up to the unconscious figure. "Dad!" He grabbed the thin shoulder and shook. "Dad?"

His dad's eyes fluttered. "Fanar?" His voice was hoarse and quiet. "Kellach... you must stop him."

Fanar grabbed him and helped him into a sitting position. "I thought you were dead." He tugged at the stubborn knot of rope securing his father. Not making any progress, he tried a different area with no gain. "I can't." Tears stung his eyes. "The knots won't come undone."

"Get Kellach. He must be stopped."

"No! I can't leave you!" It just came out; he didn't have a chance to filter it. No time to find the best way of presenting it or leading up to it. It simply exploded out of him. Fanar gazed into his dad's eyes and swallowed. "I can't leave you." Fanar felt like collapsing into himself as his dad stared at him.

Contessa came up beside Fanar with a knife and cut the ropes off. She looked between Fanar and his dad. "I found the knife in the back." She and Fanar helped his dad stand.

"We'll have to let the council know—" His dad's weak grip on Fanar loosened, and he closed his eyes.

Fanar caught him and lowered him to a sitting position. "Dad?"

"He's too weak," Contessa said.

"I can't leave him here!" he shouted. Contessa stared at him. He lowered his voice. "I'm sorry. I just can't."

"What can I do?"

Fanar sighed. "Get Oswin." She turned to leave. "And get Dr. Chernoff.

I think we can trust him." He hoped they could trust him.

He sat next to his dad. Kellach knew that Contessa had been stung by a bug. The same bug Dr. Chernoff had found? The same one that he brushed off his hand? If so, then it was fortunate he had not been stung. How did Kellach know the toxicity of the bug, and how did he know about the ground-covering plant? He obviously hadn't been helping Dr. Huff.

"Fanar?" Oswin called.

"I'm in here!"

Oswin and Rhona ran into the room. Oswin stared at Fanar's dad. "It's really him. He's alive."

Fanar gave a slight smile. "He's alive. I don't know what Kellach did to him. He's so thin, and I think his nose is broken."

"Let me see if I can find something to clean him up," Rhona said and left.

"Where's Contessa?" Fanar asked Oswin. "She's getting Dr. Chernoff."

"Guess I won't be getting through the portal today," Fanar said, more to himself than anyone.

"It still hasn't opened."

Rhona ran in and held a wet towel out to Fanar.

Fanar took the towel and wiped the blood from his dad's face, being extra careful around the nose and eyes. "No? I wonder why."

"Maybe they thought last month was the last one and they expect us to open it from our side," Oswin said.

"But we didn't have the full four hours last month. We didn't get everything we need. If we open it from this side, we only have thirty minutes. That's not enough time."

Oswin shrugged. "I don't know. That's just my idea."

Dr. Chernoff rushed into the room, Contessa on his heels, and he knelt next to Fanar's father. "James?"

"I can't get him to wake up," Fanar said. "He was awake for a little bit then passed out."

"He needs to be at the clinic. Wait here, and I'll prepare a rover." Dr. Chernoff got up and ran past the others and out the door.

Contessa sat next to Fanar and rubbed his arm. "Why do you think Kellach did those awful things?"

"I can only think of one reason," he said. "To force everyone back to Earth. He must be a Greener."

"I bet he's their inside man," Oswin said.

Dr. Chernoff appeared in the doorway. "I have it ready. Let's take him to the clinic."

# CHAPTER THIRTY

## Redemption

*This is Fanar. My dad's alive! Susan is amazing. She figured out my dad was bleeding internally and consulted the Oracle on how to perform an operation and stop the bleeding. Kellach held my dad captive for reasons we don't know yet since he hasn't woken up. We believe Kellach's also the one responsible for the sabotage and causing the lethal sickness. The portal didn't open at the scheduled time, so we're going to open it in five days after it's fully charged. Landing is in chaos. (System mark: Day 62)*

anar handed Contessa a plate of food. She had insisted on staying at the clinic with his dad while he took a break. He would never admit it, but he really needed one. Oswin and Rhona had only left a couple of hours ago. They had stayed with him as he kept watch over his dad.

"Thanks," she said. "Something Rhona made?"

He nodded. "She used some root for seasoning. I think she's getting better."

"I wish we had some salt. I never thought I would miss salt."

"We need to see what Kellach has on his term—" Movement from his dad caught his attention. "Dad?"

His father's hand twitched, but he didn't open his eyes.

Fanar grabbed his dad's hand. "I'm here, Dad." The hand squeezed Fanar's own as the monitor beeped rapidly and showed the heart rate spike from 70 to 120.

Contessa set her plate down and left the room. Moments later, she returned with Susan.

"What happened?" Susan asked.

"I don't know," Fanar answered. "His hand just started twitching." His father gasped and pulled his hand away from Fanar, his eyes wide. "Dad?"

He stared at Fanar for a couple of seconds then blinked. "Fanar?" He scanned he room. "I'm at the clinic. How did I get here? Where's Kellach?" He pushed up on his elbow, his body shaking.

Susan put her hand on his back to support him. "You're still weak. You need to regain your strength."

"We found you two days ago," Fanar said. "Kellach, he… he killed people. We don't know where he is. Why was he torturing you?" Torture seemed to be the right word. Not only did his father have a broken nose, but they had also found several bruises on his body when they'd gotten him to the clinic.

His father gave in to his frail state and lay back. "From what I could gather, he's high in the Greener hierarchy. That man who died on the day of departure, I believe he worked with Kellach. Did something to the plasma conduit so that it exploded when activated."

Fanar nodded his understanding.

"Kellach wanted my access code to the wormhole generator so he could activate it after completing what he started. Those people who died, that was just the initial test." He closed his eyes. In a whisper, he admitted, "He got the code."

Contessa put her hand to her mouth. "He's going to kill all of us."

Fanar clenched his jaw. He couldn't let Kellach get away with what he'd done. Spending all this time pretending to be their friend and secretly working against them. Not just sabotaging things, but killing too! People should not kill other people! "Not if we get him first." He looked at his dad. His father's breathing became regular as he fell into a natural sleep. "We have to." He turned and left the clinic with Contessa on his heels.

"What are we going to do?" she asked.

He talked as he walked out the door. "We have four or five days before we can open the portal, depending on how well these new panels are working. We catch Kellach and take him to the police. I want Oswin to see if he can look at the terminal and find anything Kellach stored there. We found him with some data discs; we need to find out what's on them. We also need to search the maintenance building. He's been keeping it locked, so it must be where he did his work."

"I'll get the others and meet you there." Contessa broke off from Fanar and headed down the road.

He glanced at the clinic. His dad was safe. At Landing! Ciabgan had been telling him all along where his dad was; he'd just misunderstood. Fanar shook his head and walked up to the maintenance building. The sun was rising—a day where it rose late. Inside the building's front area, he found Dr. Chernoff with a couple of his associates.

"Fanar. How's your dad?"

"Better, thank you. No sign of Kellach?"

Dr. Chernoff shook his head. "Were you able to learn anything from your dad?"

"Kellach got my dad's access code for the portal. My dad says Kellach's a leader for the Greeners and intends to kill all of us before going back to Earth."

"Then he'll be back for this. Kelly?" He waved over a young woman in her early twenties with a freckled face and red hair. She grabbed a case and handed it to Dr. Chernoff. He opened it to reveal a tiny vial of a pale yellow liquid. The vial reminded Fanar of pictures he had seen of expensive perfume. "You remember that insect I found, the one you said you saw beyond the falls?" Fanar nodded. "It has a highly toxic sting. It's probably what made your friend sick." He nodded toward the vial. "Based on a brief scan of what we found on the terminal, we believe this is the poison. A drop would kill us all. He weaponized it."

*Just a drop?* "Did you find anything else? Kellach said something about a natural antidote."

"I'm afraid not. He may have taken it with him. There's some advanced equipment in one of the rooms. You can go look."

"Okay." Fanar walked down the hall and stopped at the room he'd found his dad in. There was nothing inside except the rope they had cut off. He could only guess at how Kellach had treated him.

He moved on. The previously locked doors had been forced open. He peered into one of the larger rooms and saw what Dr. Chernoff was talking about. The room had analyzing equipment and a couple of units to create arbitrary chemical compounds. It must have been sent the previous month and secretly set up. No wonder Kellach had kept everyone out.

"What's all this?" Oswin asked from behind him.

Fanar turned to see Oswin with Rhona and Contessa. "Kellach made a poison that's so strong, Dr. Chernoff says a single drop would kill us all. He used this to make it." He looked at Contessa. "It's from the insect that made you sick."

"I never liked that man," she said.

"Tess said something about accessing the terminal," Oswin said.

"Yeah," Fanar said. "We caught him with some data discs. I'm hoping he has a copy on the terminal."

"There's one in here. I'll see if I can find anything."

"Dr. Chernoff thinks he'll be coming back, so we need to be careful." Fanar yawned.

Contessa put her hand on his arm. "It's been a long couple of days. Why don't you get some rest?"

"I want to search around back. Maybe there's something there we missed." He yawned again and walked to the large open area. Several rovers sat ready for use. The sun poured through the open bay door, making it easy to see. He wished he knew what he was looking for.

"I thought the bay doors were locked," Contessa said. His eyes snapped open. They had been, which meant—

Before he could even think it, he saw Kellach dash out the bay door. Fanar sprinted after him and quickly closed the gap. Kellach stopped and turned. Suddenly, the impact of a fist against Fanar's chest knocked him down.

Kellach bolted. Grunting, Fanar took chase again and dove at Kellach's legs, squeezing his eyes shut as they both hit the ground, the air exploding from his lungs. Kellach twisted and kicked out, but Fanar kept his arms wrapped around Kellach's legs, gasping for breath. Seconds later, fists rained down onto his shoulders and head then stopped.

"You have no idea what you're doing! I'm protecting the planet!"

Kellach roared.

Dr. Chernoff, one of his associates, and a soldier—Randy's brother, Mike—pulled Kellach off Fanar. Kellach squirmed and jerked himself about in an effort to escape.

"Don't let him go, Michael," Dr. Chernoff said as he and the soldier dragged Kellach away.

"Fanar, are you okay?" Rhona asked. With her were Contessa and Oswin, who helped him up.

"I think so." It hurt to breathe. It hurt to walk or move his arms. It hurt to think.

"We better get you to the clinic," Oswin said.

Contessa spoke quietly. "He needs some sleep. He was up all night with his dad."

"He can sleep at the clinic," suggested Rhona. "They have some cots in there."

They walked Fanar to the clinic, where Susan also suggested he get some sleep.

The room was dark except for the faint glow of equipment monitoring his father's vitals. Fanar sat up, causing the ceiling to illuminate the room at a low level. He was rested. More than he had been in a long time. His dad stirred and opened his eyes.

Fanar went to his father's side. "Dad?"

His father appeared to be so much better. Color had returned to his face, and his eyes were alert. "Fanar. You're okay." He scowled. "Kellach, he— "

"We got him, Dad. He's in custody now."

His forehead relaxed. "Good. When Earth opens the wormhole, we can turn him over to the authorities."

"You mean when we open the portal. It's past time for Earth. They never opened it."

"They didn't?"

Fanar shook his head. "No. We have three days—" He glanced at his uplink, and his eyes went wide. "I lost an entire day?" No wonder he felt so rested. "We should be able to open the portal in two days."

Dr. Monier nodded. "Good. Then we can turn Kellach over. We can coordinate with them on prioritizing Niton for the next opening and sending us fresh supplies. We really missed out last month. Barely got the one transport through. Then we—"

"Dad."

"—can get some people over to—" "Dad!"

"What?"

"Dad, we need to get you to the hospital." Fanar swallowed. "Let them

put someone else in charge until you're fully healed." His dad stared at him. "I'll be okay. I just need—" "You almost died."

"It's true," Susan said.

Fanar turned to see her standing by the door.

"Dr. Monier, Fanar's telling the truth. If he hadn't found you when he did, you would have probably died within a day. It took a while to get you stabilized. I had to consult the Oracle on how to stop your internal bleeding."

"Dr. Huff could have…" Dr. Monier's voice trailed off when Susan closed her eyes. "Oh. She was one of those who died?"

She nodded.

Fanar suggested to his father, "Dad, you can lead the team to the next planet."

"It's not the same, and you know it."

Yeah, Fanar knew it. "You can always come back here."

His father remained silent for several seconds. "A personal report of our experiences here could be useful to those preparing for the next world."

Fanar relaxed in relief. Returning to Earth would be a difficult step for his dad. "Exactly. Those log entries are useful but don't tell the whole story. Your emotions and thoughts at the time are not fully passed on." His words sounded vaguely familiar. Where had he heard them before? Fanar nodded as he continued, "We can take Kellach back to Earth, and you can get better."

# CHAPTER THIRTY-ONE

## Manhunt

**F**anar walked to the maintenance building. Dr. Chernoff and the others had vacated it after Kellach had been secured in the meeting room where Ciabgan had been held. He walked down the hall to the room where Kellach had his lab. Oswin sat at the terminal.

"Find anything?" Fanar asked.

Oswin turned. "Hey, Fanar." He gestured at the terminal. "I think there was a lot more to Kellach than we ever would have guessed. He has a full genome of that insect here."

"I didn't realize you could sequence the genome so quickly." "Not with the equipment I have. This is something special." "Anything else?"

"Just what Dr. Chernoff said. The formula for the poison and an analysis of how it attacks the body." He shook his head. "I can hardly believe it."

Fanar thought about the time he'd spent with Kellach. The man had been a little odd, but he would never have guessed Kellach could do something so awful. He felt sick.

"There's a lot here," Oswin continued. "I'll have to keep digging."

"Okay. I'm going to keep looking in the building. I didn't finish before." Fanar watched as Oswin turned back to the terminal and gave a thumbs-up. Fanar made the motion of searching, but his mind churned over the past events. Kellach had spent some time with Dr. Huff in searching for the toxin that had made Contessa sick. Something must have been found without her realizing. Anything found would have to have been transferred using data discs because no fiber had been laid between the buildings yet. Maybe that was why Kellach had taken those discs, because they had held the information needed to make more of the toxin.

Fanar stood straight, his eyes wide. There was only one way to see why Kellach needed the discs; he would ask Dr. Chernoff for them. After telling Oswin where he was going, he went to the lab at the end of B street, where Dr. Chernoff worked. Inside, dozens of small habitats lined the walls where insects were held for study. Several people inside worked at various tasks, but not Dr. Chernoff. He recognized the back of Kelly by her red hair and approached her. "Kelly?"

She turned, holding a leaf with white clusters on it in some tweezers. "Yes?"

"Do you know where Dr. Chernoff is?"

"He's with that mob trying to get answers from Kellach." Her eyes went wide. "Who'd have thought one of our own would do something like that?" She shook her head. "To think I helped him."

Fanar scowled and took a step back. "You helped him?"

She held up a hand. "Not like that. He came in here, um, maybe a week before the portal opened last time? Anyway, he held a dead insect and asked if I had already seen it. When I said no, he asked if I could make a quick examination. I made some observations about the stinger and venom gland. I asked if I could keep it, and he said no. He seemed excited about it and left." She stepped toward Fanar. "How could I know he was going to use it like that?"

"He fooled all of us. It isn't your fault. Did he say where he got the insect?"

"I asked, and he said a rover came back from the waterfall with it."

*He must have found the insect in the rover we used when Contessa fell in the river. Even then he was planning something nefarious, and that insect provided just what he needed.* "Do you know if Dr. Chernoff got some discs from Kellach?"

Kelly shook her head. "No."

Fanar thanked her and headed toward the meeting hall. As he approached the entrance, he could see what she meant. The doors were open, and people filled the building. He squirmed his way toward the front and saw Kellach sitting in the meeting room at the end of the table, a half-smile on his face, while Dr. Chernoff yelled a question at him.

Dr Yamamoto, Dr. Williams, and his dad were in there with him. *What's he doing here? He's supposed to be resting!*

"They've been at it for hours," Rhona said. "You walked right past me when you came in. Your dad just got here."

Fanar watched as Kellach sat in silence, daring the board members to break him. Giving up, they left Kellach and exited after instructing Michael to watch him. Fanar approached his dad. "What are you doing here? You need rest."

His father nodded. "I must admit, I'm tired. But I had to hear it firsthand. Kellach claims he's been acting alone, and questioning the soldiers seems to support that."

Fanar watched Dr. Chernoff walk by. "I'll be right back." He caught up to the entomologist. "Dr. Chernoff?"

The man turned, his bushy grey eyebrows standing in stark contrast to his fury-fueled red face. "Yes, Fanar?"

"Did you find any data discs on Kellach?"

"One. It was locked, but we were able to force him to unlock it." "Can Oswin look at it?"

"It's important—"

"Please. He had my dad, and Oswin's really good with chemistry, and he would be able understand anything on there. I just need to know what's on it."

Dr. Chernoff held out a storage case. "Be sure he tells us right away when he finds something."

"I will."

"May I have everyone's attention, please?" Dr. Yamamoto said over the speakers. After everyone quieted, she continued, "I know everyone is anxious for answers, but it's been a long day. We will keep Kellach in custody here and return him to Earth when we open the portal in two days. Let's get some rest."

With some grumbling and the occasional off-color remark, the crowd left the meeting hall. Fanar scanned the throng in an attempt to see his dad. Soon the only people remaining were a couple of the board members talking to Michael. Hopefully his dad had gone back to the clinic. As he turned to leave, he saw Oswin by the door, talking to Rhona.

Oswin saw him approaching. "Sounds like they didn't get anything from Kellach."

"Maybe not, but we did get this." Fanar opened the storage case and placed the single disc inside into the tray on the bottom of the case. The top showed a blue dot and a description: "Our savior."

"I'm surprised it wasn't locked."

"Dr. Chernoff said they forced him to unlock it."

Oswin nodded and accepted the case. "I'm hungry and tired. You mind if we wait until tomorrow?"

"That's fine. I gotta find my dad." Fanar left them and went to the clinic, where he found his father in bed.

"It's more comfortable than the couch," his dad said.

Fanar couldn't help but smile at that. It was almost like having his dad back. The dad he remembered from when he was a child.

Fanar woke to the emergency air horn. He rolled off his cot and asked his dad, "I don't suppose you'll stay here?"

His father pulled the blanket off himself. "Not a chance. Let's go see what the emergency is."

It was a short walk in the dark to the meeting hall. Inside, they found Susan squatting next to Michael. Dr. Chernoff stood to the side. Fanar's dad took command. "What happened?"

"Kellach escaped," Dr. Chernoff said. He pointed at Michael lying on the floor. "I came in to talk to Kellach and found him there."

If what Fanar had witnessed yesterday was any indication, Dr. Chernoff's idea of talking meant a lot of yelling. He ran to where Kellach had been held. The door stood open, and the room was empty. If he still expected to kill everyone, he would need that toxin. He ran over to where Dr. Chernoff was talking to his dad. "Dr. Chernoff, where's that bottle?"

"Bottle?"

"The toxin. Where's that bottle you showed me?"

Dr. Chernoff swore and ran toward the door. The arrival of others responding to the alarm slowed his exit. Fanar ran after him. In the meeting hall, he could hear his dad over the speaker asking if anyone had seen Kellach or the remaining soldiers.

Fanar easily caught up to the jogging Dr. Chernoff. "Where is it?" "It's in a locked drawer of my lab. I didn't know where else to put it."

Fanar left Dr. Chernoff behind and ran past Kellach's house on B street. It was just a couple of houses away from the lab at the end of the street. Inside the lab, several drawers were pulled out, their contents discarded onto the floor. Cabinet doors near the entrance hung open, the countertops below them littered with broken bottles and specimens.

Dr. Chernoff entered, breathing hard. He swore again and lumbered to the left to where a drawer had been pried open. He slammed it shut. "We need to"—he took a couple of deep breaths—"get back to the"—he gulped in another breath—"meeting hall. Start a door-by-door search."

Fanar could practically hear his own heart racing at what was to come. "How will he use it? Are we too late?" He swallowed and took a shaky breath. He was scared, pure and simple.

"I don't know. Let's not find out." Dr. Chernoff walked from the lab to the house across from Kellach's. "You'd best come in with me."

Fanar followed him inside and through the outer hall to the master bedroom, waiting as Dr. Chernoff grabbed a case from the drawer under the bed. He put his thumb onto the security lock and opened it to reveal a hand pistol.

Fanar's eyes went wide. Nobody was supposed to have guns except law enforcement. "You have a gun?"

Dr. Chernoff ignored him and walked to the wall closet to retrieve another case. He opened the second one and took out some of the ammunition. He met Fanar's gaze. "I'm sure you have a lot of questions, but we don't have time for that. I'll check Kellach's home. You meet up with your father and tell him the situation. We must find Kellach."

Fanar wasn't sure he wanted to go alone. Searching Kellach's home sounded dangerous, but being on his own... that sounded worse. He nodded. The moment they left the house, he sprinted toward the meeting hall. He looked back to see Dr. Chernoff's silhouette entering Kellach's home. Fanar hoped the entomologist would be okay and left the lit road to cut between houses, heading straight toward the meeting hall.

Groups of people in clusters of five or six were going from house to house, their hand lights projecting a dizzying display of white dots on the ground as they searched for Kellach and the remaining soldiers. Occasionally a light would blind him and then continue on its way.

Inside the meeting hall, he found his dad giving directions to a couple dozen people. When they left, Fanar approached his dad. "He got the toxin."

# CHAPTER THIRTY-TWO

## Farewell Earth

anar's father sat on a chair and closed his eyes. The only other time he'd appeared so defeated had been at his wife's funeral. Fanar stood there, unsure of what to say or do. After several seconds, his dad opened his eyes and took a deep breath. "We must make sure he doesn't use it. He can't hide for long with everyone looking." He grabbed the back of another chair and stood. "Let's get a rover and help with the search."

Fanar frowned and stepped close to him. "Are you okay?" "Yeah." His father sat down and sighed. "Maybe not."

"What's wrong? What can I do? Should I get Susan?" He couldn't let something happen to his dad. He might not have been able to help his mom, but he intended to be present for his dad.

"I just need to rest." He held Fanar's gaze. "I'll be okay."

Those three small words eased Fanar's tension. His dad had been through a lot, so Fanar could understand his fatigue. Fanar sat and stared out the window. A group of zigzagging spotlights on the ground disappeared as a search party entered another home. He hoped Kellach and those soldiers could be found soon. He thought about the gun Dr. Chernoff had. Dr. Chernoff was probably the only one who had a chance against those soldiers and the guns they carried.

"There you are." Oswin came in with Rhona and Contessa. He sat next to Fanar.

"Still no sign of him?" Fanar asked.

"No, but I'm thinking"—Oswin dug out the disc case from his pocket— "maybe we can learn something."

Fanar stared at the case. He really wanted to know the contents of that data disc. Kellach had grabbed it when he and Contessa had surprised him, so it had to be important. "I can't. I need to stay with my dad."

"We'll stay with your dad," Contessa said. "You can go."

Rhona turned her face from the doorway. "We will?" She stared at her sister, who narrowed her eyes at her. "I mean, of course we will." She plopped onto a chair and frowned at Contessa.

Contessa sat next to Fanar's dad. "You don't mind our company, do you?"

He smiled at her. "Of course not."

"You sure, Dad?" Fanar asked. "I can stay."

"I'm anxious to know what's on there too. You go."

Fanar hesitated. "Where's that soldier, Michael?" He believed that Michael could be trusted since the other soldiers seemed to have turned on him and helped Kellach escape.

"They took him to the clinic. He was hit pretty hard on the head," his father said.

Oswin jumped up. "C'mon."

That motivated Fanar to move. He stood and followed Oswin toward the door. "If you need me—"

"I'm okay, Fanar," his father said. "And I'm in good company."

Oswin ran across the field toward the maintenance building with Fanar at his heels. They went inside and stood behind the counter, where Oswin put the disc into the terminal. The last file viewed opened automatically.

"It looks like the toxin stuff you were talking about earlier," Fanar said. "Anything about an antidote? Kellach said something about the spores of the plant we found Contessa on being a natural antitoxin."

"I didn't see anything about an antitoxin." Oswin scanned through the file. "No, nothing in here about that. It might be in another file." He closed the file and opened the directory. "What's this?" He opened a file called "Delivery" and read silently.

Fanar read over his shoulder. "He sprays it as a mist?" Fanar asked aloud. He kept reading. "He's already poisoned the water that's going to Earth! He's going to vaporize it into the atmosphere!"

"No wonder he was so angry about the pump." Oswin turned to face him. "The other day, right after you left to find your dad, we had a problem with the pump. Kellach got over-the-top angry about not losing any of the water going to Earth. With that much poison, he could kill millions."

A thought struck Fanar. "Remember when the bladder was ruined and we lost all that water? I bet Kellach did that so he would have an excuse to work on it and put the poison directly into the bladder."

Oswin stared at him. "I think Dr. Huff suspected something."

"What do you mean?"

"Right before that happened, I went to the clinic to see if she would help find your dad, but she hid behind a locked door. She didn't come out until she knew it was me."

"And he killed her."

Oswin looked past Fanar, his eyes wide. "I'll take that disc now," Kellach said. Fanar turned. "How—?"

"How did I come up with a brilliant plan to save this planet and Earth? It was you, actually. When that girl got sick, something inside of me just knew that whatever made her sick contained the answer. The little beauty must have gotten caught in her clothes and crushed so I could find it the next day. Shame it had to die in such a tragic way, but its loss will be for the better good."

Fanar gawked at Kellach. Instead of the hunched-over guy with the funny way of talking, the man stood tall and spoke with an intensity Fanar had never heard before. "Why?"

"Why? I thought you would be asking to help." He continued when he didn't get a response, "I've seen the way you handle yourself. You don't like the idea of this planet being destroyed any more than I do. I know, because I saw you there at the Greener meeting. You didn't stay long, but you were there."

Oswin gasped.

Kellach resumed, "You're smart; why don't you come with me? I have a lot of the antidote left."

Antidote? Of course! The bottle didn't have the poison; it had the antidote. The water had already been poisoned.

"We can keep this planet from being desecrated, and when we get back to Earth, we can keep them from doing the same to other planets."

Fanar stood speechless. He knew of the mistakes that had led to the poor conditions on Earth. Everyone did, but this? *This is insanity. You don't kill millions of people to prove your point.* "There's a better way."

Kellach drew his lips into a straight line. "Well, then, I guess you can stay here. Though it will get rather lonely once everyone else is gone." "What are you—" Oswin started. At that moment, the emergency siren blasted from the meeting hall.

"Right on time," Kellach said. "Everyone will go to the meeting hall and soon be met by my soldiers. Goodbye." His eyes shifted, and he nodded.

Fanar turned to see Oswin hit on the head by a soldier with the butt of a gun. He turned to face Kellach but didn't duck in time to prevent being knocked out by an object hitting his own head.

Pain seared through Fanar's head. He opened his eyes and found himself inside the maintenance building with Oswin beside him. The emergency air horn blasted for five seconds before going silent. "Oswin?" He looked at his friend. Oswin breathed. Fanar shook him, and Oswin groaned and moved his hand toward his head. *Oh, good, he's going to be okay.*

A more pressing thought entered Fanar's consciousness. "Dad." He stood then grabbed the counter and leaned over it as blackness threatened to overtake him. He stood there for several seconds before he could see, and he staggered out the door.

If he could still hear the emergency horn, then he hadn't been unconscious for long. Lights and shadows of some colonists converged onto the meeting hall and into the trap. He faltered forward several steps then stopped. Should he help his dad, or should he stop Kellach? His dad's words came to him. "You do the right thing, and everything will turn out right." But what was right? As he contemplated this, the impact of the portal opening washed over him.

He looked toward where the portal would be. He hadn't thought it would be ready yet. He turned back to the meeting hall. *What should I do?* He ran his fingers through his hair and turned back toward the portal. He had to stop Kellach, but his dad… Tears ran down his cheeks.

He had to help his dad! He took a few steps toward the meeting hall and stopped. He had to stop Kellach! He stood frozen. Small spots of light from all directions slowly made their way toward the meeting hall. He turned toward the portal and with a grunt turned again toward the meeting hall. He cried and growled as he wrestled with the impossible decision. *I can't just stand here. I have to do something!* So many points of light, all heading to their deaths. How many on Earth would be killed?

He turned toward the portal then back again to the meeting hall. He sighed and ran back into the maintenance building to get a rover. *Oh, God, please let this be the right thing.* He glanced into a room as he ran toward the rovers and stopped. *It may be just what I need.* He grabbed a small box marked for excavation that contained a small brick of explosive and a remote detonator and quickly made his way to the back, where the rovers waited.

He opened the bay door and climbed inside the closest rover. He entered an activation code, but it failed. *Great!* He sprinted to the front and found Oswin sitting against the wall, holding his head. A good sign. Fanar pulled up a new activation code from the terminal and ran back to the rover. After unlocking the rover, he stomped on the accelerator and nearly hit the side of the opening as he left the building. He spun the wheel and pointed the vehicle up past the Oracle.

The grey nothingness of the portal shimmered on Fanar's left next to the building. Fifty meters away, a platform carrying a full water bladder approached the portal with Kellach piloting it. Fanar halted in front of the platform and leaped out of the rover.

Kellach stopped, got out of the pilot's cabin, and stepped to the edge. He yelled down at Fanar, "Decided to join me?"

Fanar put his hand up to avoid being blinded by the platform's lights.

"I'm not going to let you do this!" Kellach barked a laugh and turned. "I'll blow up the portal!" Fanar yelled.

That got Kellach's attention. He came back to the edge. "You want to go back to Earth more than I do! You wouldn't destroy your only way back."

Fanar opened the box and pulled out the small explosive charge and remote detonator. "Yes, I would."

Kellach held up his hand. "Take it easy, Fanar. Let me come down and let's talk about this." He crossed the platform and climbed down the ladder.

Fanar took two steps toward Kellach. "You have to stop the soldiers from killing my dad."

"Of course, of course." Kellach dropped to the ground. "You understand, don't you, Fanar? I'm just trying to do what's right. Someone has to get these people to understand."

"You have to go now. There isn't much time." Fanar glanced back as if he could see through the buildings to where his dad sat awaiting his doom.

"Why don't you give that to me, and we can fix everything."

*This isn't right. He should be leaving.* "You need to go, or I'll blow up the portal."

Kellach snorted. "No, you won't. Here's why. If you blow up the portal, you'll be stuck on Niton for the rest of your life. You remember that explosion last time you were on Earth?"

Fanar gasped. Was Kellach saying what he thought he was saying?

"See? You do understand. The portal generator isn't there anymore. It's gone. They won't be opening the portal for you in a month or even a year because we won't let them!"

"You're lying." Fanar took a step back. "There are too many people there who would stop you."

Kellach jumped at Fanar and grabbed for the detonator.

Fanar twisted just in time and sprinted toward the portal. Like never before, he poured all of his energy into making his legs go as fast as they could. His lungs and heart worked overtime to provide the oxygen to his muscles. The ground blurred past him. He chanced a glance over his shoulder but couldn't see Kellach. *Where is he?* Fanar glanced over his other shoulder and saw Kellach falling behind.

Fanar arrived at the portal and placed the explosive device on the ground a couple of meters away from the corner of the building near the open portal. He pressed the button to enable the charge. As he stood, Kellach knocked him to the ground, causing the detonator to fly out of his hand. He scrambled to his feet and raced past Kellach to retrieve the detonator. With a dive, he grabbed it and turned into a sitting position. Kellach was nearly on top of him. He flipped the protective cover of the detonator. Even in the dark, he could see Kellach's eyes go wide.

His finger hovered over the button. *What if he's telling the truth and this is the only way back to Earth?* Past the building, he could still see lights on the ground as people moved toward the maintenance building. If he did this, he could be dooming them all. Should he really do this?

Kellach reached for the detonator. Fanar closed his eyes and pressed the button.

The blast propelled Kellach forward onto Fanar, who pushed the man off and got to his feet. Smoke poured out from the hole in the wall created by the explosion. Littered on the ground were pieces of metal along with glowing coils of wire. The portal had closed, perhaps for good.

"You've ruined us!" Kellach yelled as he stood. "You've ruined Earth!" He lumbered at Fanar. "I'm going to make you pay if it's the last thing I do." He struck out with his fist, hitting Fanar square in the jaw and knocking him down. "You can be sure everyone here will die too."

Terrified, Fanar crawled backward on his hands and feet away from Kellach. A pain shot through his hand as it landed on a sharp piece of metal. He picked it up and threw it at Kellach. It twisted and flew to the side. It was nothing like a throwing blade.

"Your dad? He's already dead." Kellach slowly approached. "Your friend Oswin? He's dead. Those sweet girls? They're dead too. But first, you're going to die."

Fanar believed him. Kellach had killed several people on Niton already and planned to kill millions more on Earth. Maybe billions. He couldn't let this man get away with it. He couldn't let him kill his dad; he just couldn't! But what could he do? He tried to move farther back and hit the wall.

Kellach stomped on Fanar's gut, knocking the air out of him. "You can't win, Fanar."

Fanar rolled away from Kellach onto his hands and knees only to receive a solid kick against his ribcage, nearly lifting him up off the ground. Another kick immediately followed, and pain exploded in his gut. Kellach was making good on his threat. Fanar coughed and gagged as he got two more kicks in his gut. He collapsed onto the ground. *This is it; I failed.* He'd never found his mom, and now he couldn't stop the death of his dad.

Fanar saw Kellach reach down and pick up a long club-like piece of metal. Fanar tried to move but found it difficult. As he tried to get out of the way, his right hand came across another piece of metal. It was long and sharp like a knife. He wrapped his hand around it. The jagged edge bit into his palm, but he wouldn't let go.

"I guess I should thank you, Fanar." Kellach stood over him.

Shaking, Fanar pushed himself up into a sitting position against the wall, careful to keep the knife hidden. "Thank me?" he asked in a hoarse whisper. "Sure. You did help in your own small way. Delivering the data disc to Ian was a big help."

Each breath was like a knife in his chest. With a grunt, Fanar shifted his legs.

"I just don't want you dying thinking you were completely useless."

Kellach jerked his arms up and brought down the deathblow.

Fanar barely managed to shift his head so the metal crushed his left shoulder. The world exploded in white and pain. When he could see again, Kellach had his arms up for another blow. Fanar quickly pushed himself forward onto his knees and thrust the makeshift knife up behind Kellach's ribcage toward his heart before falling back.

Kellach looked down and dropped the club. He pulled at the metal in his gut, getting blood on his hands as he removed it. In seconds, he dropped to his knees and soon after passed out.

Panting, Fanar sat there. Watching. His whole body shook, and pain radiated everywhere. He ran his hand over his face and gawked at the figure, still not believing it to be true. When there was no movement for nearly five minutes, the reality of it sank in. *I killed him. I killed Kellach.*

He heard several pops of gunfire before passing out.

# CHAPTER THIRTY-THREE
## Sunset

This ... Fanar. I destroyed ... portal ... waiting. Again. ... I'm hoping ... soon.

(System read error)

"It's only natural to feel this way. You must remember to tell yourself that it's about survival," Dr. Williams said.

Fanar nodded. Survival. He wasn't sure if he'd accomplished that. After his encounter with Kellach, a handful of people had found him by the damaged portal. Susan had saved his life. Now he was recovering from a broken clavicle, several broken ribs, and a ruptured spleen. He had learned later that Mike Thompson, Randy's brother, had gone into the meeting hall and shot the other soldiers. A retaliation for their role in getting his brother killed in Kellach's scheme.

Many people avoided Fanar. When he approached a group, they would cease talking or disperse. He knew it was because he had blown up the portal. The undirected blast had damaged both plasma conduits and a portion of the building. Others welcomed him as a hero because he had stopped Kellach from reaching Earth.

This is survival?

"How are you sleeping?" Dr. Williams asked. "Okay."

Dr. Williams sighed. "I'm here to help you. You endured a terrible ordeal. Lying only hurts yourself, so I'll ask again. How are you sleeping?" Fanar made eye contact. The first time he had in the session. "I'm not. I can't. Every time I close my eyes, I see him torturing my dad, or I see him killing someone, or I see…" He fell silent and looked away, unable to continue.

"Fanar, you can say it."

He glanced at Dr. Williams then stared at the floor. He blinked the moisture away from his eyes.

"This is a safe place. You can say whatever's on your mind."

Fanar looked up again. "I see myself killing him. Every time I close my eyes, I see myself killing Kellach. You happy?" I'm not.

"No, but it's important to talk about it. I know it's painful—"

"How?" Rage erupted from the depths of Fanar's soul. He leaned forward. "How can you know that every time I think of Kellach, I think about how he plotted all along to kill us? How he tortured my dad and threatened my friends and how I killed him? I killed him! I didn't want to, but I did! I killed him, and I see his blood on my hands every time I close my eyes!" He had killed someone, the worst thing any human could do to another.

Dr. Williams sat quietly for several seconds. "I'm sorry you went through what you did. I can never know, truly, how you feel because I didn't experience it. I'm here to help you sort through your feelings and to find a way for you to close your eyes and see beyond the most recent events. Do you understand that?"

Fanar nodded and sat back, wiping the tears away. "I'm going to give you something to help you sleep." "I don't want—"

"You need to sleep, Fanar. Promise me you'll take them."

Fanar held out his hand to accept the pills, but Dr. Williams didn't hand them over.

"I need to hear you promise."

Fanar rolled his eyes. "I promise." He accepted the pills, got up, and carefully walked to the door. Before leaving, he turned. "I'm sorry for yelling at you."

"You weren't yelling at me; you were just expressing your frustration at a person willing to listen. Can I expect to be a willing listener tomorrow?" Fanar nodded and left. There were a dozen things around Landing that needed his attention. The water pump that wasn't fully repaired. The extent of damage to the portal generator that had to be established so materials could be found and, somehow, repairs made. Some systems in housing units and work buildings that needed to be diagnosed. Work that needed his attention, but right now, he just wanted some time alone.

"Fanar?" his father called out and made his way to Fanar. "How are you?" After getting a small shrug, he put his hand on Fanar's arm. "We'll get through this. Together."

Fanar gingerly embraced his dad as tears ran down his cheeks. At least he hadn't lost his dad. He held on for several seconds before letting go.

His dad gave Fanar a tender smile. "I'm going to see how they're doing with the water pump. I thought maybe you could help fix it. That's your thing, right?"

It *was* his thing, and his dad finally accepted it. "Okay."

# Epilogue

Ciabgan stood in the cave behind the waterfall. It had been several sun cycles since Fanar had left in a fit of anger and anguish, and Ciabgan believed it necessary to visit him. He was steeling himself for the pain of becoming invisible when a figure plodded up beside him.

The Trociabk looked like him except for a gold tint of the outer skin. "The leader needs you," she said. Having delivered her message, she selected a rock from the floor and ate it.

Lost in his own thoughts, he thanked her. Some of the Trociabk might doubt Fanar's authenticity, but Ciabgan didn't. He couldn't. Somehow, he and Fanar were entwined in a fate, set by Aiyam, that could not be avoided. He wound his way through the tunnels until he found his father in conference with the Elder and another Entrusted.

"The suppressor is not functioning," the leader said. "The key crystal was removed."

Ciabgan knew the suppressor protected them from the Boshag. A thought occurred to him. "You believe Shanar has taken it?"

"No, we must accept the possibility that the Boshag have gotten influence over one of our own kind to prompt such an act."

The elder leaned his head in toward Ciabgan. "We must find the crystal. We need you to assist."

* * *

The story continues...

AARON L BRATCHER

SONG OF THE BOSHAG

GALAXIES
COLLIDE

# Author's note

Thank you for reading *Galaxies Collide*. I hope you enjoyed the first book of this new series.

As an independent author, reviews are very important. I would be abundantly grateful if you'd take five minutes to leave a review on Amazon and/or Goodreads. It would go a long way in supporting the continuation of this story.

https://www.amazon.com/Galaxies-Collide-Aaron-L- Bratcher/dp/1088094082

https://www.goodreads.com/book/show/149955560-galaxies-collide

# About the author

Aaron L Bratcher, a multi-genre author, has written several books to cater to diverse interests. When he's not crafting stories, he works as a Senior iOS Developer at JP Morgan Chase. Outside of work, he cherishes time with his family and friends, engages in travel adventures, or volunteers at his local church. Aaron also leads a local writing critique group and finds joy in playing games.

Visit his website at https://AaronLBratcher.com to learn more about his work. You can also connect with him on social media at @AaronLBratcher on Facebook, Instagram, and X.

Find more books at https://AaronLBratcher.com